To Mike + Trisa Ichu
With special thanks for
your friendship.
Dick Wenzel
30 June 2005

Stalking Microbes

A Relentless Pursuit of Infection Control

By

Richard P. Wenzel

authorHOUSE™

1663 Liberty Drive, Suite 200
Bloomington, Indiana 47403
(800) 839-8640
www.AuthorHouse.com

First published by AuthorHouse 05/06/05

ISBN: 1-4208-2006-0 (sc)
ISBN: 1-4208-2005-2 (dj)

Library of Congress Control Number: 2004195076

Printed in the United States of America
Bloomington, Indiana

This book is printed on acid-free paper.

To JoGail

My companion in life's orbit, always charting the coordinates of love and offering the time and space to explore and the gravity to remain close by.

PREFACE

I am one of those fortunate people who knew from age eight that he wanted to be a doctor. Like other physicians before me, as a child I had been a patient hospitalized for several weeks and became enamored with medicine since that time. When people speak of a calling or a passion for one's work, I can understand that immediately. We who have had this driving focus on our life's work find ourselves saying to young people, "find what you love, your passion, whatever it is and pursue it." And we cannot easily imagine how others might say that they go to work to earn a living so that they can pursue their hobbies and avocations in their free time. I have been pursuing my avocation most of the waking hours of my entire life.

As an infectious diseases specialist, I went to college for four years followed by medical school for four years, spent four additional years training in Internal Medicine and Infectious Diseases, and later another year studying to get a Master Degree in Epidemiology. My career interests have included life threatening infections, especially those that occur in hospitals, and I have worked to prevent and control individual infections as well as infectious diseases epidemics. Infectious diseases epidemiologists can practice medicine at the bedside, one-on-one with an ill patient and also examine populations of patients with epidemiological and statistical tools to determine risk factors for infection and their serious outcomes. The simultaneous perspectives on sick people and sick populations are broad and complementary. The two overlapping disciplines of clinical medicine and clinical epidemiology offer rich insights to infectious diseases, their victims, the microbial causes, and life itself. That is the backdrop for this book.

The essays herein describe the interaction of individuals and specific microbes, each with their unique history, personality and reactions to the interaction - what we call infection. Various perspectives are linked together by two key themes that I address continually. The two themes that weave throughout the book are the crux of the essays. The first deals with special characteristics of critical importance for a physician: curiosity and independence. When I speak of curiosity I mean the passion and willingness to ask questions, to inquire no matter what unfamiliar or controversial territory those questions might lead one to enter. And when I address independence, I mean a term that implies an element of isolation, the doctor's need to rely on himself/herself regardless of the cost in courage, social isolation or uncertainty.

The second theme relates to how one might characterize the practice of good medicine. More specifically, what is the essence of medicine? To me, the graduating physician has little ability to characterize medicine, much less what it should look like. Only with time and experience can one gain increasingly incisive ideas of medical care. My own perspectives have evolved with continual exposure to various aspects of our profession but have been especially colored by the experiences I describe in the eight essays.

The specific perspectives include the following: the reaction to being a patient, learning to be a physician, the meaning of individual and group identities among human study volunteers, the process of discovering a new disease, the drama of uncovering the causes of deaths in ICU patients from infection, the role of surveillance in the control of infection, the need for clinicians to make quick and correct diagnoses, and the importance of detective work to uncover the causes of obscure infectious diseases. Throughout the experiences that I describe, I was at a different part of my own life and career, yet I offer the lessons of these experiences from both remote and current perspectives.

Each essay focuses on a specific microbe or several pathogens, and a personal experience, before and after becoming a doctor specializing in infectious diseases.

The first three chapters deal with various themes of preparing for a career in medicine and the last five chapters with specific themes as a professional. Of course I am trying to expose the reader to my search for the ideals of our field, the activities that describe the best of our traditions. That search still continues for me.

Richard P. Wenzel, M.D., M.Sc.
Richmond, Virginia

ACKNOWLEDGEMENTS

My family has always indulged my passion for Medicine and in this case my writing the series of essays enclosed. My wife, JoGail, on many occasions read various versions of the text, patiently offering insights and asking important questions. My daughter Amy, son Richard and son-in-law Eric Volles have offered critique and encouragement. A special thanks to Kenny Marotta, a writer and freelance editor, who on many occasions in various versions offered ways to clarify my ideas. He was continually encouraging and helpful. The manuscript was not final until I made several revisions suggested by an expert editor, Joel Pulliam. Thanks also to three people who assisted in the preparation of the manuscript: Wanda Bates, Lisa Bundy and Barbara Briley.

TABLE OF CONTENTS

PREFACE vii

ACKNOWLEDGEMENTS ix

PART I PREPARATION 1

    THE PRINCE OF PATHOGENS (Staphylococcus aureus).............. 3

    K.G. (Vibrio cholerae).. 15

    THE "CORPS" (Mycoplasma pneumoniae) 33

PART II PROFESSIONAL LIFE... 49

    DISCOVERY (Rickettsia).. 51

    DEATH IN THE ICU (Serratia marcescens) 65

    IN THE SHADOW OF SEMMELWEIS (Enterococcus, Proteus
    mirabilis, Streptococcus pyogenes)................................. 81

    THE ORIGIN OF FEAR (Neisseria meningitidis) 99

    A BUMP IN THE NIGHT (Bacteroides, Klebsiella pneumoniae,
    Plasmodium vivax, Coccidioides)....................................111

EPILOGUE 121

READINGS 125

INDEX 135

PART I

PREPARATION

THE PRINCE OF PATHOGENS
(Staphylococcus aureus)

It was the first day of the summer recess after third grade from Saint Madeline Sophie School in Philadelphia's Germantown section. Engaged in one of my favorite pastimes, I was scrambling up the limbs of an old oak tree in the back yard of my pal Jackie, who lived directly across from our small house on Hortter Street. We both preferred the perspective of the world from our high perch, where above the scrutiny of parents we could freely trade opinions about school, current cowboy and indian movies, and baseball. From about 20 feet up I recall hearing the sound of a branch give way and immediately finding myself on the ground, sensing instantly that I shouldn't attempt to move. My dad was called to my side, and a red paddy wagon doubling as a police ambulance was summoned to transport me to Germantown Hospital's emergency room. The sharp pain in my right thigh was later diagnosed as a sign of a fractured femur.

There was little cushioning from the clapping of the tires over the brick and cobblestone streets along the way, and I was relieved to arrive at the comfort of a motionless hospital stretcher. After a visit to the Radiology suite for an x-ray, the fracture was confirmed, and a medical decision was made to place a Kirschner wire surgically through my thigh to assist in positioning the two fragments of bone for good healing. With a fresh hospital gown in place I was then ushered to the operating table, where a doctor placed a rubber cup over my nose and mouth, instructing me to inhale the pungent ether.

Awakening from anesthesia, I was captive to the cables, pulleys and weights designed to immobilize and align my right femur. It appeared that I had become fastened within a Jungle Jim of white pipes anchoring the stays and my injured leg in place. More than a little disheartening for the first day of vacation, the transition from freedom and mobility was stunning to an eight-year-old.

In the late 1940's one could expect a two to three week stay in the hospital from my fracture, but I was destined to be there for over six weeks. Though I remember my surgeon's visiting intermittently and briefly during that time, I can easily recall the frequent contacts with the attentive nursing staff all in starched white dresses with even more highly starched and uniquely styled nursing caps, indicating the specific school from which each had graduated.

In the third week of my stay I developed fever, felt sick all over, and noticed that the area on the inside of my thigh near the entrance of the

Kirschner wire was hot, painful and tender, and red. A small volume of sticky yellow pus oozing from around the wire on to my thigh was sent to the microbiology laboratory for culture.

My fever and malaise seemed worse in the afternoons and evenings. Despite a high temperature, I felt cold, the result of the body's shivering attempt to dissipate excess heat. I wanted covers during these times, and I felt miserable. Fastened on my back, I couldn't even turn over.

The lab report came back the day after the culture was taken. An infection was responsible for my symptoms, including the fever. Every morning the higher blood concentrations of the adrenal hormone cortisone lowers the body's thermostat, whereas by evenings our temperature rises to 98.6°, "normal," or even a little higher. Infection exaggerates these usual fluctuations. The specific infection I had was *Staphylococcus aureus,* "Staph infection."

Bacteria are classified by genus and species. The genus Staphylococcus, with many etymological links to Greek language and culture, means "cluster of grapes." The organism was given its name in the early 1880s by Sir Alexander Ogston, Professor of Surgery at Aberdeen in Scotland. Seeking the cause of pus occurring in wounds, he visualized these organisms through the microscope, by using a purple stain of aniline violet solution, applying this solution initially on the discharge from the leg wound of a young male patient. Although he didn't invent the procedure, he extended the concept of staining bacteria placed in use a few years earlier by the German physician, Robert Koch, who would later discover the bacterial cause of tuberculosis. Before publishing his paper, "Micrococcus Poisoning," Ogston had carefully examined the pus from over 100 abscesses he had studied. He noted that sometimes the bacteria arranged themselves in chains (Streptococci) and other times in indigo-colored clumps to which he gave a name derived from the Greek language, Staphylococci. The species name, *aureus*, derives from the Latin word for gold, a reference to the yellowish hue the tiny colonies of Staph have when grown on the surface of the gelatin (agar) culture medium.

Intact skin is a major defense against such infections, providing an impervious barrier to potentially aggressive organisms. However, the presence of the foreign body in my thigh – the wire holding my bone in place – provided a transcutaneous bridge and allowed organisms to gain entrance to the underlying soft tissue.

The Kirschner wire may also have allowed fewer bacterial organisms to lead to an infection than would be needed by trauma alone. A key experiment reported in 1957 by two surgeons, Elek and Cohen, demonstrated the importance of even suture material in lowering the number of bacteria necessary to cause infection, the infecting dose. These two physicians performed experiments that would not meet muster in ethical terms today but were sanctioned in the 1950s. They found that if they made shallow incisions in the arms of young male volunteers, no infection occurred until the infecting dose was escalated to one million Staphylococci. However, if they left a suture in place for a brief time, the infecting dose was as low as 100 organisms – a 4-log reduction from 10^6 to 10^2 *S.aureus*! The infecting dose for my infection may have been well below a million organisms.

The Staph causing my infection were multiplying in colonies just below the skin. Furthermore, having traversed the surface of the skin, bacteria have great avidity for attaching themselves to any available wire or plastic, like barnacles on the surface of a boat beneath the water. The bacteria then secrete a complex sugar substance, a biofilm or "slime," that surrounds their colonies within a clear protective shield, walling off an assault by neutrophils, the white blood cells that are the body's major defenders against infection. This can happen with all foreign bodies placed in the body, including prosthetic joints, vascular catheters, and orthopedic wires.

When various metal or plastic prostheses become infected, in general it is necessary to remove them in order to provide effective antibiotic treatment. Curiously, the organisms attach themselves in layers to the foreign material, and the layers closest to the plastic contain bacteria that multiply very slowly and tend to ignore any circulating antibiotics that were prescribed. This means that most antibiotics that interrupt the formation of cell walls (penicillin drugs) and most of those that interfere with active protein development (such as erythromycin) may not be effective, since the bacterial metabolism is slowed so much. Those bacteria in the farthest out layers – away from the plastic – multiply at faster rates and as a result are more susceptible to antibiotics. They are not so much of a problem for antibiotics. But once the germs get a foothold, it is too late for antibiotics alone, and the foreign material needs to be removed.

In 1948, only the intramuscular form of penicillin was available, and I received injections into my buttocks at six-hour intervals around the clock for three weeks. The Kirschner rod and the billions of bacteria attached to it were removed early to reduce the microbial burden in my body and to assist penicillin in curing my infection. My buttocks became pincushions, tattooed with dark reddish-brown specs at the sites where the injections were given. Nevertheless, after several days I began to improve, but convalescence was slow, and my expected stay was extended significantly.

Whose Staphylococcus caused my infection? Obviously, hand contamination of the exit site of the wire had occurred, possibly by one of the medical team or possibly from me the patient. This organism lives in the moist mucous membranes lining the inside of the nose, and a third of all people carry Staph in a type of peaceful coexistence, usually with no harm resulting. My vantage point currently would suggest that there is some possibility that it was my own *Staphylococcus aureus* originally residing in the microscopic condominiums of my nasal passages and carried to the skin near the wound and wire. Although it may not be obvious to many, we live as individuals and have evolved as species with billions of bacteria, especially those living in the upper respiratory and lower gastrointestinal tract and those on the skin and scalp. In fact our cells number 10^{13} – a 1 followed by 13 zeros - and the number of bacteria colonizing our cells number 10^{14}. As unpalatable as it may seem, we people are 10 parts bacteria to one part human!

More likely, if instead the Staph causing my infection came from a health care worker who had not washed his or her hands before touching my wound site, it probably originated in the wound of another post-operative patient on the ward. After all, Staph are the most frequent cause of infections after surgery. Alternatively, it might have originated in a nurse or physician who had had some skin disease, perhaps a mild dermatitis, a minor Staph infection such as a stye, or just in a nasal carrier who coincidentally had a large number of Staph bacteria on the hands. Forty percent of people who harbor *Staphylococcus* in the nose have the exact species on their hands.

Years later on the faculty at the University of Iowa Medical School I had the opportunity to study one of the topical antibiotic creams available that kills *Staphylococcus aureus* and eradicates the organism from the nose. When we applied the antibiotic to the nares of volunteers twice a day for five days, over 95% of the carriers lost their *S. aureus* in the nose. Of interest, in the same volunteers, the Staph could no longer be found on the hands. One cannot escape the conclusion about how Staph gets to the hands of carriers. My colleagues and I approached this question earlier in a paper that we published in *The New England Journal of Medicine*. In 1973, Drs. Owen Hendley, Jack Gwaltney and I reported on the number of times individuals touch their nose with their fingers. In that study we, in fact, were examining the possibility that if a common cold virus were placed on the fingers – for example, after shaking hands with a person with the Rhinovirus infection – how likely is self inoculation to the nose. In Sunday school, the figure was 1 time per 21 person-hours of observation. In medical education seminars it was strikingly higher: 1 per 2.4 person-hours! So there is probably an ongoing traffic of organism transfer in both directions between the nose and the fingers.

In recent studies when applied to the nares of hemodialysis patients in Europe, the topical antibiotic not only reduced the rates of nasal carriage of Staph in these patients with kidney failure but also lowered the rate of true infections, including life-threatening bloodstream infections. Previously, elegant studies in the 1960s showed that patients undergoing common surgical procedures who carried *Staph aureus* in their nares had two to three times the rate of post operative wound infection with Staph than non-carriers. Knowing these two facts, I wondered that if we could eradicate nasal carriage of patients pre-operatively, perhaps the corresponding rates of incisional wound infection post operatively would be reduced. In 2002, our team showed that when the antibacterial ointment was applied to the nares of patients about to undergo surgery, the rates of *Staph aureus* infections were reduced after surgery – but only among those who were nasal carriers of Staph. Studies such as these confirm the fact that many of the staphylococcal organisms causing infection arise from the patients themselves, and – even if uncommonly – the carriage of Staph can lead to an infection.

Some organisms are equal opportunity pathogens showing no bias in host preference. Staph respects neither class nor station, and infections occur from the ghetto to the White House. Consider this: shortly after George Washington became President after our nation's independence, he developed a dangerous carbuncle on his thigh. The most likely cause of this soft tissue infection – as with my own childhood infection – was *Staph aureus*. Seeing his suffering and illness, Washington's aides became acutely worried, and an eminent New York surgeon, Dr. Samuel Bard, was summoned to treat the infection surgically. Fortunately, the drainage of the abscess was successful, the President would be cured without the advantage of antibiotics, and he would successfully complete his term.

From what we know over 200 years later, it is reasonable to conclude that the father of our country was almost certainly a nasal carrier of Staphylococci. Washington died in December 1799 of what most experts think was acute bacterial epiglottitis causing upper airway strangulation. Of course no one knows the cause of his infection, but a prominent organism that can do this is *Staphylococcus aureus*.

There were no medical experts in Infectious Diseases in 1948 when I first encountered Staph. The specialty would have to await the proliferation of antibiotics, their various indications, doses and side effects. There was also no formal concept of infection control, although I suspect that out of self-interest many nurses washed their hands after managing my wound dressings. I just cannot recall hand washing in the hospital. If knowledge in this discipline had been more highly developed, the next 13 years of my life might have taken a different direction.

A year after my hospital discharge, I again developed fever, pain, redness and tenderness of the inner side of my swollen right thigh. All of these are signs of inflammation. I had two round scars, each an inch in diameter, one on the inside of my thigh and one on the outside, corresponding to the exit sites of the Kirschner wire placed surgically during the hospitalization. A small abscess had developed at the site of the inside scar. A local physician in Germantown lanced it, and I again received some intramuscular penicillin. Although the physician thought that he saw a tiny speck of bone in the pus, in fact I never developed infection of the bone, osteomyelitis. Fortunately for me, penicillin again lysed the cell walls of the bacteria growing beneath my skin, and I was well again.

When I was in 6th grade we moved from the Mount Airy part of Germantown to Wyndmoor, just outside of the Philadelphia city limits. My new family physician lived just across the street from me, his office was five blocks away, and I was one half block from my new school, Seven Dolors, the seven sorrows of Our Lady. One to two times a year my chronic infection would flare up with all the signs of inflammation, disrupting grade school, my high school classes and football season at Chestnut Hill Academy, and college activities at Haverford. Each time my physician with his great smile and confidence brought out the syringe full of penicillin, and each time I would feel better within days. What I did have for another thirteen years was a recurrent sinus tract infection caused by *Staph aureus*. The organisms had lodged themselves in the relatively avascular area of the scar made in response to the track traversed by the Kirschner wire. Thus, the bacteria were somewhat isolated and protected by the poor blood perfusion of the scar tissue. Periodically – about once or twice a year – the body's defense would be outwitted by the Staph, and I would be sick again.

In retrospect, despite his warm "bedside manner" our family physician apparently had modest curiosity, and I was a victim of his lackluster approach. He met only two of three features that some experts think are essential for a physician: he was "accessible" and "affable", but not completely "able". He failed to recognize the ingenuity of *Staph aureus* to hide like a fugitive, a prince of darkness in a protected corner of the body, and the fact that a cure awaited excision of the sinus tract, a surgical procedure that would occur when I was a first year medical student at Jefferson Medical College in Philadelphia. Dr. Anthony De Palma was the Chairman of Orthopedics, an internationally recognized surgeon who had written separate books on large joints, including the knee. When my infection recurred, I asked him for help and advice, and he recommended surgery, essentially the removal of the tissue giving safe harbor to Staph. I agreed to have the operation over the Christmas holiday hoping that this plan would prevent my missing school or in any way jeopardizing my grades.

In December 1961, I entered a hospital again, this time in the downtown section of 11th and Walnut Street, Jefferson's academic hospital. The operation was brief, the scar tissue and bacterial remnants excised, and the wound was packed with gauze that would be withdrawn slowly over days so that the tissue's healing with a rich vascular supply would occur from the inside out. However, my anticipated quick departure would be delayed, and again due to *Staphylococcus aureus* – again hospital acquired.

Was this infection unconquerable? Once, it would have literally been so. In the pre-antibiotic era, a revealing study of *Staphylococcus aureus* bloodstream infections published in 1941 by Skinner and Keefer showed an 82% mortality. I was, however, the beneficiary of the emerging antibiotic era. In fact, back in 1948, I had been among the first civilians in the United States to receive penicillin, a marvelous antibiotic that saved tens of thousands of lives of military personnel during World War II. It also was beneficial in the social repatriation of our fighting force. In 1945 post war Germany, Dr. Ben Kean was the Army medical officer in charge of administration of all penicillin in that country. As he pointed out in his 1990 book, *MD*, the ostensibly enigmatic Army "decreed that Penicillin in Germany would be reserved for two privileged groups: American solders and German prostitutes!" The general in charge, General Morrison Clay Strayer, responding to the strident criticism of the German civilians that the wonder drug was not also available to them flatly stated, "we plan to send our boys home clean!" Clearly in its inaugural years, penicillin enjoyed a broad spectrum of activity against a wide variety of pathogens.

In 1961, however, my hospital-acquired infection was resistant to penicillin. In the 13 years since my first encounter, the vast majority of Staph in the United States had developed – really been selected for – resistance to the wonder drug, penicillin. In most colonies of bacteria, there may be one in a million or one in 10 million with resistance to an antibiotic. Early in the use of antibiotics, this is of no consequence since the vastly susceptible population of bacteria is eliminated. However, with prolonged exposure – selection pressure – the initially rare, resistant species became dominant. Now, millions of bacteria have the genes coding for antibiotic resistance, and at this point the antibiotic will not work to cure the infection. In the early 1960's, U.S. hospitals were riddled with outbreaks due to penicillin-resistant Staph. Fortunately, the gods were good to me because two years previously the drug methicillin had been developed to combat penicillin-resistant strains. It had a slightly different structure chemically, and the bacterial enzyme (penicillinase) that Staph mobilizes to break up one of the carbon bonds in penicillin was not effective against drugs like methicillin. An oral analog soon became available, and I knew this because Dr. Frank Sweeney, an expert in Infectious Diseases on the faculty at Jefferson, prescribed the new experimental drug for me. I had not appreciated it then, but the specialty of Infectious Diseases had also emerged since my original hospital stay in 1948. Clinical research on anti-infectives had also quietly arrived and, although I had not signed a human volunteer consent form and had not classified myself as a "study subject," in fact that is exactly what I was! As a result of the new antibiotic and the surgery I never had a recurrence of my chronic sinus tract infection. A competent surgeon, Dr. De Palma, an expert in infectious diseases, Dr. Sweeny, and a new antibiotic had conspired to cure me of the Staphylococcus.

One part of the journey I had started in 1948 ended; but another part continues. Staph was a constant companion in my younger years, invisible to the naked eye but erupting into periodic infections like the fits of an unpredictable volcano. What that meant for me was a series of interruptions from school, athletics, and social life while waiting for antibiotics to quell the microbial disturbance. Before my surgery in medical school, I had assumed that I would have this problem all of my life. No previous encounter with medicine led me to think otherwise. However, *Staph aureus* sparked the idea that doctors should be insatiably curious about the causes of disease, not just their temporary control. The former requires energy and tenacity, whereas the latter assumes only expediency. Maybe there are available and lasting cures, and the periodic dousing of smaller fires is not the best that medicine has to offer.

When I was back in the hospital at 21, I wasn't too old or too mature to experience the emotional deprivation that hospital treatment can bring. Specifically, because of the post-operative infection with penicillin-resistant Staph, I was isolated. At the entrance to my room was a basin for handwashing and cloth towels for drying. On visiting me, not only the medical and nursing staff but also my mother and father had to don gown and mask, and I at 21 was acutely embarrassed by the circumstances of a new staphylococcal encounter. It was the patient-not the organism-that was isolated I thought, punished for the sin of being infected with a bacterial pathogen. I was sad, I felt guilty, and there was no confessional for relief.

Years later as a physician, I would reflect on this experience and the fate of similarly isolated patients. Being in the hospital is an alien experience in and of itself. Consider these words of authors and nurses, Woods and Edwards, on the emotions of being a patient: "the basic human desires, hopes, fears, the image one has of oneself, the vitality of the ego are all suddenly at stake. These feelings of lack of control, the loss of self..... are components in serious illness. By having to turn to others, to give up certain rights, to become passive, accept the decision of others..... we adopt the role of patient." Now add to this the sudden closure of the door, the subsequent requirement for all physicians, nurses, orderlies, food service personnel, janitors, friends and family members to be enveloped with mask, gown and gloves. What does a patient think? Worry about? Does anyone in the medical profession take the time to say that none of these precautions offers a single advantage to the patient? That it is an infection control effort, an attempt to contain the microbe within the tight quarters of the hospital room? Does any physician or nurse ask a patient what it feels like to be placed in "isolation"? Does any health care professional admit that the patient is a victim, not the cause of a problem? As a result of this experience I could contrast the emotions of a distraught patient in an unfamiliar environment with the feelings of comfort instilled by medical professionals. The doctor can make a difference, and a doctor doesn't have to be Captain Marvel or Superman to make people feel better. But he does have to care about the patient.

The summer of 1948 was a defining period in my life – my first encounter with a hospital, surgery and Staph. I became an unwilling patient in a medical system with significant shortcomings. Even today in the 21st century, there are important challenges remaining with *Staphylococcus aureus*. Staph cause most of the skin and soft tissue infections occurring in the community. They are an important cause of bacterial pneumonia complicating viral infections from influenza. In the hospital Staph are the number one cause of post-operative wound infections, the second most frequent cause of bloodstream infections, and an important cause of pneumonia acquired in intensive care units. Staph not only have a special ability to acquire genes that program the cell to resist the effects of antibiotics but also to acquire individual genes that target the skin to cause a rash, the gastrointestinal tract to cause diarrhea, and the blood vessels to relax, and lose all tone leading to hypotension – the toxic shock syndrome. This syndrome was initially apparent in tampon users whose *Staph aureus* became trapped in the vaginal tract. Later, ear-nose-throat patients showed up in shock after nasal packing, done to control nose bleeds. They too suffered from toxic shock due to the Staph in their noses. And even today some with minor post-operative wound infections from Staph develop life threatening shock because the organisms harbor lethal genes within.

Many questions remain: Why are some people subject to bacterial colonization and infection and others not? Why do some colonized patients become infected and other do not? When infection erupts, why does it do so at that specific time? Can we find new ways to control antibiotic resistance in Staph? This is an especially important current question because in the summer of 2002, a *Staph aureus* strain was isolated from a woman in Detroit that was resistant not only to penicillin and methicillin, but also – for the first time ever – to our stalwart drug, vancomycin – previously always available as a clutch hitter, when striking out was not an option. A second vancomycin-resistant Staph infection occurred in a patient in Pennsylvania a few months later. Both patients had had prior infections with methicillin-resistant *Staph aureus* and had received weeks of vancomycin before the uniquely resistant strain emerged. A third case occurred in New York state in 2004. A fourth case was reported from Detroit in 2005. Since the substrate for the development of vancomycin-resistant Staph seems to be the presence of methicillin-resistant Staph (MRSA), it is important to know that MRSA comprise 50% of all Staph infections in U.S. hospitals. The reason that the figure is so high is primarily due to poor infection control, especially poor compliance with handwashing by medical personnel in U.S. hospitals. Our low level of hand hygiene is no match for the genetic repertoire of *Staph aureus*.

While infection control experts ponder some of these issues related to patient care, it is important to emphasize the tenuous advantage of antibiotics over modern strains of Staphylococci. Consider this: the earth is 4 billion years old, bacteria have evolved for over 3.5 billion years and *Homo sapiens* has evolved for only 500 thousand years. We are newcomers, children compared to bacteria in life experiences and the ability to adapt. And the antibiotic era is barely 60 years old. Most importantly, fifteen years ago there were over 10 pharmaceutical companies making antibiotics, and now only a hand full show some interest in providing new drugs to treat infections. The problem is the consequence of a business decision, perhaps compounded by increasing barriers perceived by industry to rapid approval by the FDA.

Large pharmaceutical industries such as Pfizer, Merck, GlaxoSmith Klein and others view expenditures for new drugs in terms of a return on their current investment. In economic terms this is the "net present value" of their expenditure, which is defined as what tomorrow's cash flow would be worth today. However, if one considers blockbuster drugs such as those for arthritis or neurological diseases or cancer, with net present values

– in millions - of $1,150, $720 and $300 , it is much lower for antibiotics – only $100.

The business decision focuses on what industry thinks they need in future income to get excited about investing in new drugs. Some might say that pharmaceutical companies are greedy and selfish, but whatever the motives of industry, we physicians and our patients are caught in the middle. There needs to be an urgent discourse involving government agencies, infectious diseases experts, and industry to solve this acute problem. In the footrace between man and microbe, our ancient bacterial cousins on the evolutionary tree are poised to make new strides. We need to react with management plans for both prevention and treatment.

In the meantime, we in medicine need to explore important questions: Are simpler therapies around the corner? Can we develop an effective Staph vaccine? Can we invest in an effective strategy for *prevention* of this serious infection? Furthermore, how can we in academic medicine select medical students who will retain a sense of curiosity throughout their demanding careers, who will continue to ask questions about their patients who remain uncured? Equally important, how do we in the medical profession select students who will care deeply about their patients? Can we really teach genuine compassion in Medical School? Specifically, what experiences in Medical School instill proper skills, empathy, inquisitiveness, and decision making? From a broader perspective, can we as physicians improve the human condition, make a difference in our careers?

Fifty years after my initial encounter with Staphylococci, I still reflect on this interaction between man and microbe. *Staph aureus* is a frequently seen pathogen, a killer organism capable of acquiring genes coding for antibiotic resistance, toxic shock syndrome, and the ability to colonize humans. These bacteria can be carried readily on the hands of people, hitch-hiking from person to person in hospitals, prisons, day care centers, barracks, and families. Eventually they can cause serious infections for which we have fewer drugs in the pipeline to treat. For those of us knights in the pursuit of infection control, the conquering of this single pathogen is the Holy Grail we seek.

13

K.G.
(Vibrio cholerae)

Ken Goodner, Professor and Chairman of Microbiology at Jefferson Medical College in Philadelphia, was an eccentric iconoclast whose passion for teaching was unparalleled at the institution. In the early 1960's, he was one of the world's leading experts on cholera, a devastating cause of diarrhea in the distant lands of India and Asia. Surrounding himself with a team of animated teachers who shared the excitement not only of imparting knowledge but of stimulating students to go beyond the text and to explore unassigned material, he taught not just the antics of microbes but also the lives and thinking of great men and women who tried to understand bacterial biochemistry and physiology. Importantly, he recounted various stories of the interaction of microbes with people----- what we call infection.

Goodner signed all notes with his initials and so was referred to as "K.G." He could be intimidating if a student failed to sense the warm humors of his personality emanating from a wonderful heart inside, but importantly he promoted free thinking at a time when most medical school faculty had a Marine Corps approach to learning: "Drop and give me 50 causes of fever!"

Goodner and his colleagues were the first to teach us the Gram stain test to identify bacteria. Even today a basic staining procedure named for a creative 19th century microbiologist, Christian Gram, is used routinely. The staining method, used for over 100 years, is a simple laboratory procedure: one can take a swab of the pus, roll the swab onto the surface of a glass slide, perform the four steps of the Gram stain, and look at the slide through the microscope. Organisms are either rod shaped or spherical (cocci), and appear either blue (Gram positive) or red (Gram negative) with the stain. The use of two different colors to classify bacteria was a major step beyond the staining procedure used earlier by Ogston.

With Gram's procedure the blue color resulting from the first stain applied, crystal violet, binds to the bacterial cell wall only after iodine is added, and it fails to fade after an alcohol decolorizer is added. In the fourth step, a dye called safranin is added. Organisms that lose the blue color after the alcohol is added now retain the red stain. The blue Gram positive bacteria include organisms such as Staphylococcus (Staph) and Streptococcus (Strep) that infect the skin, and the red Gram negative bacteria include those such a *E. coli*, Salmonella and Shigella that infect the lining of the intestine. The whole staining procedure takes only a few minutes but can give important clues to the likely organism causing infection.

The morning lecture in Microbiology was followed by a laboratory session, at which time a quiz – an unknown Gram stain to challenge us students – lay in waiting on the individual microscopes. Goodner had a playful yet creative side, and occasionally he would instruct his team of younger colleagues to place the slide upside down on the platform of each microscope. Unless a student realized that there might be a reason why the bacteria never came into sharp focus, he or she might not consider pulling the slide out, checking to see if it was placed rightside up and rotating it properly under the lens. He encouraged his wards to question basic assumptions.

I met K.G. in the second year of medical school as one of his students, about the time when I was developing a jaundiced view of medical education. Up to that time it was an oppressive experience, often based not on the strength of data or the force of logic but on the dead weight of authority and the heavy burden of imparting existing biases. The challenge was not the complexity of concepts, but the huge volume of material we students were expected to absorb like human sponges. When the exams came, we simply squeezed our brains onto the answer sheet.

In college the emphasis for those majoring in biology was biochemistry. The tradition of teaching at Haverford, one that I attempt to pass on to my own students, was based on the history of ideas, an understanding of the key question at the time of inquiry, of scientific hypothesis, experiment, the acquisition of new information, and a remolding of the existing model. I didn't mind memorizing the molecular structure of the four bases of DNA: adenine, guanine, cytosine, and thymidine. There was a story there, any three of the four providing the code for the building blocks of aminoacids, of their assembly to form proteins, the key enzymes that orchestrate vital biochemical tasks and bring harmony to the body's performance on a macro level. But no professor in the *first* year of medical school offered the history of the rich science of anatomy, physiology, histology or biochemistry. That richness would have to wait until the second year.

Furthermore, in Medical School we had not had any encounters with patients yet. In the 1960s, the first two years of medical education were called the "pre-clinical" years because the focus was classroom and laboratory based science. Unfortunately, there had not even been any clinical scenarios presented to relate the science with medical care. It all seemed so abstract.

In his formal presentations K.G. made all the difference in my thinking, and a few additional experiences cemented my respect and affection for

16

him. The afternoons in the Microbiology class began with small group sessions led by a faculty member usually with 15 students to a group. We were grilled with questions about microbiology. Frequently, however, the inquisition had nothing to do with our lessons because KG wanted to observe our responses to unfamiliar situations. One day he asked about the value of mouthwashes as antiseptics and pointed a commanding finger at me. I hadn't the slightest idea of data to support an opinion but responded sheepishly, "Dr. Goodner, there must be something to this idea because one of the products kills bacteria on contact, the result of 'Magic GL-70'."

Of course, I was alluding to frequent radio and television advertisements about a popular mouthwash in 1962. My real goal was to avoid hesitation, to respond quickly, hoping that classroom wit would save me, yet I suddenly realized I was taking a serious chance with one of the country's leading scientists. For a brief second he seemed incredulous that one of his best students would cite such nonsense, but K.G. had a comforting lift of his eyebrows and announced that the group would advance to the next topic. Of course there were no compelling data that any mouthwash offered anything more than normal oral hygiene, the brushing of teeth. If I had known the literature, that would have been the correct answer. But that wasn't the point. The issue was how does a student physician handle uncertainty? Years later I would recall and increasingly appreciate his extraordinary tolerance on that afternoon.

During and after World War II, K.G. had been a frequent traveler to corners of the globe that I had barely heard of: Madagascar, Malaysia, Morocco and others. Diseases of these exotic lands were well known to him, and he developed a keen interest in the biology of cholera. He brought not only science but also images and challenges to the didactic lessons of microbiology, and I could imagine the suffering of victims of infections even though I had not yet cared for a single patient.

In classes he described a typical victim of this Gram negative bacterium, *Vibrio cholerae*: dehydration from massive diarrhea so severe as to cause the eyes to sink deeply into the sockets, and the skin to lose all turgor, remaining ruffled in folds for seconds after being pinched by the fingers. Patients were listless and had a dull facial appearance. Their breathing was somewhat rapid, the result of acid buildup in the blood from loss of alkaline bicarbonate in the stool. The body was attempting to blow off acidic carbon dioxide (CO_2) to restore proper chemistry. Of special interest, because the organism does not invade tissue and cause inflammation, the patient has no fever. This is in striking contrast to so many other non-cholera diarrheas such as Salmonella, Shigella, and Campylobacter.

Something else occurred to me: good physicians treating victims of infectious diseases might make an important difference in the lives of many patients. It was obvious from K.G.'s presentations that hundreds if not thousands of people were sometimes involved in epidemics of infections.

After two years of basic science courses, the clinical years began in the third year of medical school, and I was thrilled to have real patient contact. We junior students spent four to eight weeks on each rotation: Internal Medicine, Surgery, Psychiatry, Obstetrics, and others to provide a broad background in Medicine and expose us to various subspecialties. Subsequently, there was opportunity to choose a number of electives in the fourth year of school, and many of my classmates would choose to visit centers where they had hoped to do an internship and specialty training – the residency.

Near the end of my third year it was rumored that K. G. was thinking of sending two students to Asia to work in a lab in Taipei run by members of a U.S. Navy Research Unit. One of my classmates heard that I was on the short list. Immediately, I ran down the six flights of the building I was in, crossed Walnut Street and raced up the four flights to K.G.'s office. "You're looking short of breath, Wenzel," he said. K.G. was alone in the far side of his lab, his silhouette reminiscent of Alfred Hitchcock's.

For years he had had a scientific interest in classifying different bacterial strains of the same species by mixing them with specific antibody and noting the clumping of organisms. To visualize the clumping – indicating a "match" or specific type – he wore horn-rimmed low-power magnifying glasses with thick lenses. When walking around the lab, he stationed these lenses above his forehead.

K.G. was wearing his "goggles" in the resting position when I arrived, and he reacted as if he had no idea why I was there. I responded that I had heard about the possibility of flying to Asia, and I was very much interested. "Oh, Asia," he said, with the same emotion one would have on answering, what time is it? "Would you be willing to travel out of the country for three months?" he asked. This time an intensity captured his voice, and he had the look of an angler who had just hooked a giant salmon.

"Yes, sir, I would," I responded without hesitation.

"Bob Phillips is a friend of mine," he continued, "the Commanding Officer of a research unit in Taipei, and he is willing to have two students join the Navy's laboratory doing research on cholera. I think that you would learn a great deal."

"I would like to go," I reiterated quickly.

In my mind it was the opportunity to explore some new frontier, more exotic than any I had previously experienced. After college I had spent two months with fellow graduates touring Europe by car, lodging in inexpensive youth hostels and delighting in witnessing new cultures, trying new food and drinking beers never sold in the United States. Otherwise I had been in only several states in the northeast U.S. In contrast, Goodner and colleagues had recounted tales of overseas experiences in jungles, deserts and mountains and the various interactions of people, their cultures and indigenous microbes. In vivid lectures they had inoculated us students with the need to explore. As a result I was eager to experience foreign travel.

"You would be representing Jefferson and me if you go. Do you understand this responsibility?" he added seriously.

I continued to nod affirmatively and said, "Yes Sir, I do. I know that I can work very hard."

On the spot K.G. agreed to send me to Taiwan, but with a key proviso: Bill Wood, a friend and classmate and I would get his permission and financial support to go, but we had to spend a month in K.G.'s lab gaining microbiological skills so that we would be useful in the Navy's Medical Research Unit Number Two, NAMRU – II. If we accepted this condition, he would organize the trip and arrange to cover the costs of the three-month visit. At the time I had no idea what basic laboratory training we would receive or whether this would be fun. But I recognized the provision as a necessary hurdle to an international adventure.

Bill Wood and I met again with K.G., and he outlined our course of instruction. The highlight of the first night was the ceremonial receipt of our personal magnifying glasses – our own "K.G. goggles." In mocking fashion our fellow classmates would rib us about this milestone in our education as we began the journey with *Vibrio cholerae*.

We learned to identify this curved, rod-shaped organism by its red appearance on the Gram stain. Vibrio comes from the Latin-to vibrate- and these organisms are motile by virtue of a single flagellum, the vibrating oar located at one end. We would also learn to type the three main families of this Asian scourge - Ogawa, Inaba and Hikojima – by adding specific antisera that caused clumping of the bacterial colonies as seen under low power magnification. We studied its biochemistry and what was known of its physiology. Most of this extracurricular learning was carried out in the late afternoons and evenings, and we prepared for our first visit to Asia.

It was the fall of 1964 when we arrived in Taipei. Our boss was Navy Captain Robert Phillips, a renowned physician and biochemist who earlier in his career had measured the exact electrolyte composition – sodium, chloride, potassium, and bicarbonate – of cholera diarrhea and designed precise formulas of intravenous salt solution for rehydration. Bill Wood and I unpacked and prepared to settle in. However, Phillips told us that in two days we would leave for Manila on a military transport plane with several Corpsmen and be there for an unspecified time. Our job was to assist the research team studying infection in cholera patients. A huge epidemic had swooped down on the Philippines, and we would very likely manage many patients on the wards, essentially keeping the victims alive with appropriate intravenous fluids while detailed physiological studies on their condition were overseen by Navy scientists and performed by the Corpsmen.

We stayed at the Manila Bay Hotel, and after breakfast spent 10 to 12 hours a day at the San Lazaro Hospital, a brief jitney bus ride away. The heat and humidity were oppressive, but I never felt more free in my life than in Manila. My responsibilities during the day were totally of my own choice, helping people survive and recover from an exotic disease. The more I worked, the more I influenced the health of the cholera victims. I also had no responsibilities after work except to explore the uncharted excitement of life in Manila.

My main task on the wards was to replace the diarrhea loss with IV fluids. First, I would tie a rubber tourniquet around the arm of the patient, seeking to identify a puffed up vein on the surface of the skin which I would gently wipe with an alcohol soaked patch of gauze. Then into the vein I would insert a steel needle connected to IV tubing. The IV bottles hanging from a pole by the bedside contained a solution containing mostly sodium and chloride but also potassium and bicarbonate, all of the electrolytes lost from the gastrointestinal tract. Once the needle was safely lodged in the vein, I would anchor its hub securely with thin strips of adhesive tape, first wrapping under and then crisscrossing the tape above the hub to stick to the skin on the opposite sides. The tourniquet was released, and the fluid began to restore the patient's critical loss of volume.

The hospital admitted 100 victims of acute cholera a day, and I began to think that after my first week I could insert an IV needle into a microscopic thread. I felt useful, and lives were being saved. The Philippines presented exotic disease management in true life patients in contrast to the interesting but abstract lessons in Philadelphia.

I learned that patients with cholera would often lose 20 liters of fluid because the organism manufactures a toxin, essentially a poison that instructs the cells lining the gastrointestinal tract to leak out plasma. The resulting "diarrhea" is in fact mostly composed of salt water with flecks of protein, and in Asia is referred to as "rice water" stool. It has a musty smell, but no foul odor. Patients die if they cannot be rehydrated with appropriate intravenous fluids, essentially replacing a liter in for each liter lost.

In poor countries where cholera is prevalent, patients are placed on canvas cholera cots, their exposed buttocks over the large round hole in center of the cot above a bucket which collects the choleric discharge. With this system the nurses can manage the care of a large number of patients on an open ward, essentially using a calibrated dip stick, measuring the quantity of fluids lost into the bucket and manipulating the IV speed to match the loss. In reality, cholera is an "instant man" package: just add salt and water, and quickly you have a person again

Here we were on the other side of the world, a Dodge City equivalent with gun-toting men everywhere at the hotel, in restaurants, and on the sidewalks. The streets were crowded with people speaking mostly the local dialect, Tagalog, but also Spanish and English, with many on bicycles and jeeps, and horns were honking continuously. We worked hard and played hard, every night the entire team going to a new restaurant, sampling exciting cuisine, often a mixture of local food with mild spices and coconut or very European style food usually with a Spanish flair. Importantly, the exotic Philippino brown skinned women were beautiful, and any religious or cultural inhibitions were instantly abandoned.

We ate grilled squid sold along the bay by individual merchants. It was cut into irregular pieces about six by eight inches, pierced by a thin wooden stick and meant to be washed down with San Miguel beer. However, the specialty of Manila was a balut, and they were hawked by young men carrying tens of each of the fertilized duck eggs in a canvas bag with a shoulder strap. The vendors would sound out their wares with a crescendo musical announcement, "baluuut! baluuut!" Two of the Navy Corpsman described the proper protocol for eating a balut: you crack off the top of the shell, imbibe the juices and save the "best" for last --- the cartilaginous mid-term embryonic chick. There were stories of gagging, retching, sometimes worse after eating a balut. Despite the nutritional value and even with several downed San Miguels, Bill Wood and I demurred.

Walking along the main streets near the hotel we would see many locals with white cotton masks tied around their nose and mouth as if on an isolation ward at home. Others tied white handkerchiefs around their faces like "bandits" in a U.S. movie. These scenes informed everyone of the arrival of a cholera epidemic.

When cholera hits poor countries, many people don the white cotton masks or handkerchiefs over their nose and mouth in an attempt to reduce the risk of infection. Cholera is spread by contaminated food or water, but the tradition of wearing masks dates from the miasma theory of disease, a medieval idea that bad air – mal aria – is the cause of disease. In Manila in

1964, many unschooled people wore masks or handkerchiefs, especially family members of the diarrhea victims.

At the hospital our day to day leader was a Navy Lieutenant Commander only ten years our senior, Ward Bullock. Ward was tall and stiff in appearance and seemed to measure his words carefully. In contrast to the easy going Corpsmen, he was quite formal, opinionated and less accessible to us students. Ward was a bright, fully trained infectious diseases specialist, but to our sensibilities much too serious a physician and scientist. Despite our naïve judgements, he would emerge as an outstanding teacher of science and we learned a great deal from him. Ward would subsequently have a prominent career in academic medicine studying the immune system and diseases like tuberculosis, deep fungal infections and leprosy.

After we had been working at San Lazaro Hospital for about two weeks, Ward announced to Bill and me that he would escort a team of four health advisors for the Philippines through the patient areas to demonstrate to them how we treat the disease and to explain the modern views of how the Vibrio is transmitted. Bill and I thought that Ward should be coaxed to be more relaxed, and when the quartet of visiting dignitaries stepped onto our wards, we were prepared. To Ward's amazement and dismay, his two medical students from the advanced scientific bastion of the U.S. were both wearing cotton masks while managing the care of cholera patients!

Recovering quickly from the momentary embarrassment of our youthful pranks, Ward explained the ridiculous situation to our guests, but no one seemed amused. We did hear about our insubordination later, but fortunately Ward forgave our rebellious expressions of independence and held no grudges. Years later, our paths would cross intermittently at national meetings, and he and I would reminisce about our shared experiences in the Philippines and chuckle about our perspectives at the time.

Weekends in Manila were free, and the Navy Corpsmen and their two medical student charges visited beaches, went to the mountains or rented a large boat to party on Manila Bay. The Philippine Coast Guard would take exception to our racing around the large ships anchored in the harbor, rock music blaring while we danced with our dates on deck. But the experienced Corpsmen would offer the Guard members a few bottles of San Miguel beer, and we were off again, only to repeat the ritual in two or three hours.

In Philadelphia I had dated fair skinned debutantes from Chestnut Hill and the Main line, daughters of conservative families. The goals of many were to marry the son of a family friend, return to the neighborhood and raise children who would attend the same schools. In Manila, Fe, my coppertoned girlfriend was worldly, and despite having limited advantages in life, she had an easy smile and was a survivor. When a typhoon passed near the island of Luzon that month, she took me to her home outside of Manila and showed me where the water was still two feet deep on the first floor, all of the important furniture and keepsakes moved upstairs. Flooding was common with these frequent storms, and periodic adjusting became a way of life. She and her family had been witnesses also to a series of epidemics, cholera being just one cause. She calmly explained this while brushing her long black hair in front of a mirror as she opened up her favorite music box and played its light melody. What I considered devastating, she took in stride.

An amusing feat that made her famous in our group of Americans was her ability to open the cap of the bottle of San Miguel beer with her teeth. This innocent appearing and lithe young woman would place the neck of the beer bottle in the side of her mouth, give a single thrust, and off came the cap in seconds. Like a character from a scene in "South Pacific," my exotic girlfriend was unlike anyone I had ever met before.

When we were on duty during the week, the medical team was at the hospital early, where we began our jobs of treating severely dehydrated patients. In babies the intravenous needle was placed into a scalp vein. In older children and adults, we looked for a vein in the forearm or on the inside of the elbow – the antecubital fossa - where large veins traverse near on surface. I was learning how to assess the degree of dehydration and estimate the speed by which to rehydrate with IV fluids. It was in Manila when I found complete harmony working with infectious diseases: I had responsibility, I was learning both technical and cognitive skills, and I could observe successful outcomes quickly. I was energized, and each day was eagerly anticipated.

By contrast, the return to Taiwan six weeks later was somewhat of a let-down because we had no clinical involvement there, and the people in the labs recognized that we were students, not scientists. A single month of afternoons and evenings in K.G.'s lab failed to give us sufficient skills of value to investigators at the research bench. We could do some menial tasks to help, but there were no lives at risk for us to respond.

The fascination with Chinese culture helped greatly to keep us animated, and we had time to tour not only Taipei but nearby towns by car. I began reading about Chinese traditional Medicine, the Red Book of Mao which I had picked up in Hong Kong on the way over, the ancient writings of Laotze and modern works of Lin Yutang. These philosophical writings explored a basic question, how should a person lead his life? It was a brief period of reflection, of contrasts and awakening. Away from home and the demands of school, I had moments to think.

During those moments, I reminisced about the last night in Manila saying goodbye to my girlfriend. What had begun as a detached romance had evolved ineluctably and unexpectedly over six weeks to a more intense relationship. Tears had welled up, the overflow of emotions dammed up by the absence of words to define and express everything that I was feeling. She, the pragmatist, knew that we would never see each other again and told me that I would be fine in Taipei, "You will find a Chinese girlfriend." I knew that that was not the point.

To me she was also the personification of the millions of disadvantaged people in the world, any one of whom could get cholera or another life-threatening infection never seen at home in the U.S. She was a special person with family and friends and someone who loved her. As a survivor in life, she was also a source of optimism. Her beauty was in part the fact that she remained positive and proud of her culture, accepted life's challenges without fear, and sought no pity. A young woman a world away from my home taught me lessons never offered in medical school. National statistics – births, deaths, infection, life spans – are the sum of real people, each with beliefs, aspiration and hopes, attempting to cope with day to day existence.

In retrospect, there were two barriers for my not having this perspective after examining patients in Philadelphia: there was so much information to memorize in medical school that there was no time to think or to reflect. Secondly, I never met a clinical role model on the wards who could offer this aspect of medical care at Jefferson.

Compared to the experience in Southeast Asia, the rest of my fourth year would be subdued. I now knew that I was destined to help people with severe infections, and I had found a passion for travel to exotic places. Perhaps I could eventually blend the two notions. K.G. was proud of his students' accomplishments, and joined by our families had met the plane on our return to the Philadelphia airport.

After medical school I elected to do an internship at the Philadelphia General Hospital (PGH) primarily because I felt that I needed more hands-on experience. The large city hospital saw every disease in the book, often at late stages, because poor people wait long times before seeking help. At Jefferson, although I had seen a number of sick patients, I never had primary responsibility. However, PGH gave me more than I ever could have imagined in responsibility and much less supervision than I would have wished for. Nevertheless, the camaraderie among the 107 PGH interns who came from all over the United States was intense, no doubt related to our need for help from each other with extremely ill patients, especially in the middle of the night. Although we didn't say it, there were times when we were fearful of managing so many new admissions with little or no faculty and senior resident oversight. The year was very important to me, because I learned that I wanted a more formal education in Internal Medicine followed by training in Infectious Diseases.

I kept in contact with K.G., and he suggested that I visit his friend and colleague in Baltimore, Ted Woodward. It was the fall of 1965 when I met Dr. Theodore Woodward, Chairman of the Department of Internal Medicine at the University of Maryland School of Medicine, a world class clinician and expert in Infectious Diseases. Woodward was a Chesapeake Bay fishing buddy of Goodner's and also a fellow traveler to exotic stations of the world whenever outbreaks of infection surfaced.

I called to request an appointment with Dr. Woodward and was astonished when he instructed me to get off the train in Baltimore and take a cab to the Oyster Bay Restaurant. We would have lunch together. I had had interviews for residency training at academic centers at Boston University and Columbia University in New York but they were formal, even rigid at Boston where I was essentially grilled for 45 minutes. By contrast, Woodward was warm and charming when I met him at the restaurant, and he talked about important changes in the program at Maryland: wards were no longer segregated. I was caught off guard by the statement because I hadn't considered that it could be any other way. Never before had I traveled south in the United States, and I had obviously failed to appreciate first hand what magnificent social revolutions were taking place. But now I knew where the Mason Dixon line was drawn.

We walked a half-mile back to his office, and he said that if I wanted a residency in Internal Medicine there, he would allow me the option of taking an elective rotation during the second year of training to spend some time in what was then called East Pakistan. I accepted his offer and in July of 1966, I moved to Baltimore for four more years of training in Internal Medicine and Infectious Diseases.

Previously, I hadn't realized how small the world was or that there might be a type of Medical Mafia until the second year of my residency. Woodward said that his friend, Captain Bob Phillips, who was also a friend of Goodner's, had retired from the Navy and was appointed the new director of a laboratory and hospital run by the Southeast Asia Treaty Organization (SEATO) in Dacca, East Pakistan. Phillips knew me from Taipei and Manila and invited me to oversee the clinical activities of the cholera and diarrhea ward for a three-month period if I wished. I signed up immediately, purchased a map of the world and looked up East Pakistan on the Bay of Bengal. The same sense of adventure that I had experienced in the Philippines had welled up again as I prepared to embark on another journey: the Indian Subcontinent.

K.G. was to retire in 1967 and do consulting for several agencies. He wrote to say that he would meet me in Dacca and offered advice on how I should travel: spend some time in Athens and visit the Parthenon, fly to Karachi in West Pakistan and then connect with a flight going overland across India to Dacca. He wrote to me that he was comfortable, had looked forward to some rest after so many years in academic medicine, and eagerly anticipated seeing me in Dacca.

Sadly, he never made it, the victim of a massive rupture of an undiagnosed abdominal aortic aneurysm. He died quickly from rapid blood loss in the small town of Pittsburg, Kansas, where he had grown up.

In a letter regarding our travel plans, he wrote to me on August 2nd 1967. He said that he was "glad that there are still eight weeks. I had not realized the amount of fatigue which could be stored up in 21 years. Things are going better now although June projects have become September projects. Weeds cover the garden, trees are barely alive, the dog has sharp teeth, the roses are gorgeous and there are many splashes of color. Details unfortunate, overall effect entirely charming. Every weekend there are ever so many slow moving cars. It's a relaxed atmosphere. Sunrise is greeted. Sunset is admired. The birdsong in the early night is pure delight. I like to watch the cows grazing in the nearby fields. Best of all I am at peace. Of course I am looking forward to the fall adventure. It is always good to be surrounded by friends whom one trusts completely."

Framed and hanging in my office at home is K.G.'s last letter to me.

In Athens, en route to Karachi and later to Dacca I had the same feeling of freedom that I experienced earlier in Manila. However, I was travelling alone, a situation that is often not as enjoyable as sharing new places with a good friend or family member. From Athens to Karachi I quickly sensed that I was the only Westerner on an Air France plane filled with Chinese communists wearing Mao jackets and pins. I felt not only different, but odd and insecure.

When we touched down in Cairo, I read a notice stamped with large block letters on my passport stating boldly that I was not permitted to be in Cairo. My anxiety heightened further when the flight took off in complete darkness, apparently as a precaution, the result of tensions between Israel and the Arab world in 1967. The Six Day War broke out on June 5th 1967, but in the subsequent fall of that year little was resolved in the eyes of the former combatants. Nevertheless, the trip was uneventful, and our plane landed in Karachi at 11:00 p.m. The connecting flight to Dacca would not board until 6 a.m., a period of time that seemed interminable with my lack of sleep, lack of language speaking ability, lack of understanding of Urdu or the script, and obvious physical differences from those milling about in the airport.

After the long flight across India, I arrived in Dacca, which was hot and humid, packed with people who were mostly poor and in the midst of a cholera epidemic. Few roads were paved, and red colored pedicabs were everywhere, the high-pitched sounds of bicycle ringers constantly filling the air. Most people were thin, and many were barefoot. The streets were congested with brown-skinned men who were bare chested and wore plaid cotton skirts tied at their waists. Nearby, thousands of poor people and their families lived along the banks of the Ganges River in order to be close to water, and some lived on crowded boats. But the water became contaminated with Vibrios once or twice a year, sometimes associated with the annual monsoons, flooding, and the migration of infected boatmen. Along with dried fish and dahl, the main dietary staple was rice, and it was customary to pour water from the Ganges River over a huge container of rice at night to keep it moist. With the warm ambient air temperature, I could only imagine the exponential growth of bacteria achieved by the next day. I was again a foreigner in a strange new land.

At the hospital I was comfortable directing a ward of cholera patients, conducting clinical rounds as we did at the University of Maryland, teaching ideas of Western Medicine and learning a new culture. The guideline in general for initially hydrating a patient with cholera was to weigh the victim on admission and run in IV fluids equivalent to 10% of their body weight as quickly as possible. A 50-kilogram man would be infused with 5 liters of IV solutions within a few hours, a situation that could cause serious fluid buildup of the lungs or even death in a non-cholera patient.

One day I was measuring the volume of fluid in the bucket of a newly admitted patient. The humidity was breathtakingly high, and my head was glistening with sweat. Suddenly while bending over, my black horn rimmed glasses slipped off my face and fell directly into a two-liter volume of rice water stool, sinking to the bottom of the plastic container. For a few seconds I just gazed into the bucket. Taken aback, my initial thought was, "Hell, I see well enough without glasses" and briefly considered discarding spectacles and rice water stool as one. But I regained my composure and recalled the fact that it takes ingesting a billion organisms to cause cholera. If I could wash my glasses and hands assiduously, I would avoid infection. I retrieved both my glasses and common sense, washed cholera organisms carefully from rim, glass, and hands, and had no risk of infection.

After a month at Dacca I was asked to board a plane from Dacca and travel close to the Burmese boarder where a small missionary hospital, hours from the coastal city of Cox's Bazar on the bay of Bengal, was experiencing a serious cholera outbreak. In contrast to the strong lab support in the U.S. or even in Dacca, the hospital at Malumghat had only an EKG machine with which we could estimate dangerously low potassium levels, and only by noting the changes recorded in the electrical conduction of heart muscle. With severe diarrhea, there can be a marked loss of potassium, important for the integrity of all muscle cells. As we carefully repleted the loss of potassium, we could notice improvement in the EKG tracings.

Phillips had introduced me to the useful copper sulfate bottles. These were a graduated series of 25 pale blue-colored copper sulfate solutions within small rectangular shaped bottles with triangular necks. The fluid of each one in succession had an increasing specific gravity. A patient with cholera would have his blood drawn, the red cells would be spun down in a centrifuge, and a few milliliters of red cell-free plasma taken up into a small nose dropper as the pressed bulb was relaxed. A drop of plasma was then squeezed into a bottle labeled "1.035." If the drop sank to the bottom, the plasma was more dense than the copper sulfate solution, and the patient was very severely dehydrated. I would try "1.040" next. If instead the initial drop has risen to the top of the solution, the patient was less severely dehydrated, and we could next try the test in a bottle labeled "1.030." The process was repeated with interpolations until the drop seemed to level off half way down in the solution. If the drop of plasma fell to mid level in any one of the bottles, that was the specific gravity of his plasma. Normal was "1.025," and a level of "1.035" meant that a patient was very dehydrated, at least 10% body weight too dry.

We not only took care of patients at the 50-bed field hospital but occasionally would travel by station wagon, raft and a few miles by foot to reach remote villages where cholera was reported. On one such trip which took about six hours, after arriving at a village of thatched roof huts, vintage 16th century, I was asked to remain in place for a brief period so that a local man could run to his nearby village and bring his family back to look at me. No one in that region had ever seen a white person before. Racial tensions characterized the U.S., Martin Luther King would be assassinated in 1968 in Memphis, and my city of Baltimore would then be burning in frustration and anger. But I was the outsider, a minority of one in this remote dot on the map of what is now Bangladesh.

I was invited into a hut where the entrance was low to the ground, and once inside, a lantern gave the only light to the 15 people huddled around the dying 8-year-old cholera victim. She was flaccid with somewhat rapid but shallow breathing. Her brown eyes had regressed deeply into the bony sockets from fluid loss. She did not appear frightened, just resigned to her fate. Her skin turgor was poor, losing all elasticity and remaining in lumpy folds when pinched lightly. She was so dehydrated that she could not talk, too weak and too dry to utter a single word. To my chagrin I could find no vein to instill the IV fluid that my translator and I had carried from the field hospital. Instead, I felt for the thready pulse of the femoral artery in the groin and placed a needle connected to an IV bag directly into the large femoral vein just medial to the artery and ran the fluid in at full speed. I repeated the procedure on the other side, both IVs going at the same time. Two liters infused rapidly in only 20 minutes, and I could now discern the billowing of veins in the arm. I started an IV line in each forearm, adding two more liters in the next 40 minutes, and the patient began to speak and look revived.

I don't know what the people in that hut were thinking: a young light-skinned man with blonde hair and blue colored eyes had put needles into a dying girl's groin, and she regained her voice and spirit. For the most part they remained subdued. Of course they were pleased, but I sensed that the main emotion was amazement. We would eventually transport her by foot, raft and station wagon for further treatment, and she left our hospital well a few days later.

I was ecstatic that I could help save a life, and a beautiful young girl would return to her caring family and village. I was also pleased that I was resourceful enough to insert IV fluids into a femoral vein. At a time of uncertainty I had an answer. I wish I could have told K.G. the story of that special day.

On the way out of the village I was asked if I would examine the young girl's grandmother who was also thought to have had cholera. But the older women had a rigidly stiff neck and a depressed level of consciousness. Her skin was burning hot; she had fever. She obviously had meningitis, a life-threatening infection of the spinal fluid and the thin layer of tissue – the meninges – surrounding the fluid and protecting the spinal cord. Hours later at the hospital our stain of the spinal fluid showed Gram positive diplococci suggesting pneumococcal meningitis, and we administered intravenous penicillin. Fortunately, she too recovered and returned to her home.

This success was a small miracle. Another day without therapy and this elderly woman would surely have died. Although any reasonable clinician could differentiate meningitis from cholera, I had that opportunity. I succeeded, and I felt great. A grandmother and grandchild would be reunited after two unrelated but life-threatening infections.

I had left that remote village the same as I entered, a stranger with no name but with locally unusual skills. I know that during my three-month visit I helped only a tiny fraction of the ill in Bangladesh. But to a few I had made a difference. That's the real treasure of medicine: it is a privilege to help other people. We should thank our patients for offering us opportunities to assist them in their quest for health.

So much happened in such a short period in the fall of 1967. Life and death and even return to life. Cholera, the organism that introduced me to tropical medicine, travel, and exotic cultures was also a light and mirror for the social revolution going on in the United States. The perspective I had gained on racial issues was unforgettable. I saw the beginning of desegregation in the South in the mid-1960s in hospitals. I witnessed the tension and anger that exploded in Baltimore and other urban areas when Martin Luther King was assinated, and I traveled from a country where I was one of the majority to another land where I was clearly a member of a small minority.

Most importantly, it is amazing for me to realize that one person, a uniquely special teacher and mentor, K.G., had such a profound influence on my life and my career. With great generosity he gave me the experience of acting on my own, an education about the challenges and rewards of self-reliance. Perhaps isolation is a part of self-reliance, and my separation from my Filipino girlfriend, my global travels alone, my isolation in the Bengali cholera village, and K.G.'s ultimate lesson – the isolation from K.G. himself as a result of his death – had been a part of the lesson of independence.

With extraordinary experiences in life, new perspectives and new questions surface. The outline of the qualities needed to be an excellent physician began to appear like the silhouette of a distant figure emerging from a fog in early morning. However, how does medicine cope with sick individuals and sick populations at the same time? How does one learn and maintain reverence for patients who are vastly different – culturally and socially – from their physician? Can we find an ideal balance between the science and the art of our craft, and can we always keep in sharp focus the individual while trying to reverse the fascinating biochemical disease processes that brought the patient to our attention? Furthermore, how does one create new knowledge that can be translated from the laboratory to the bedside for the benefit of sick patients? How do I train young physicians in the best conduct of clinical studies? As a Chairman of a Department of

Internal Medicine overseeing the training of 130 interns and residents each year, I continue to explore these questions and refine my responses.

THE "CORPS"
(Mycoplasma pneumoniae)

It was in the Spring of 1966, with my internship year at Philadelphia coming to a close, when a few of my fellow trainees received alarming notices that in July of that year they should convey their combat boots and Army issued uniforms to designated stations around the United States for initial training. Many would later be assigned to battalions that were fighting in dangerous military zones with exotic names like Da Nang, Mekong Delta, and Cam Ranh Bay. This was common for those arriving directly out of internship without specialty training.

All young physicians in that era had to join the military draft, and initially we were presented with a list of services and invited to order our preferences. I chose the Navy because my dad had served on destroyer escorts in World War II and had spoken highly of his Naval experiences. I also asked and received permission from the Navy to be deferred until my training in Internal Medicine and Infectious Diseases was completed. Physicians could do this under the Berry plan of deferment because the military needed a smaller group of physicians in specialties like Internal Medicine, Surgery, Orthopedics, Psychiatry and others. The Berry plan was established in 1954 and named for Frank B. Berry, M.D., who instituted the plan while serving as Assistant Secretary of Defense (Health and Medical).

A still smaller group with subspecialty training in Infectious Diseases, Trauma Surgery, Post Traumatic Stress and others was needed during War. My interests surely coincided with those of the military because one of the major causes of disease and sometimes a leading cause of death in wars is infection, and the so-called Viet Nam "conflict" proved to be no exception.

My military service would be postponed for four years of further training in Internal Medicine and Infectious Diseases at the University of Maryland in Baltimore. During that time, because I was eager to learn about viruses and viral culturing techniques, Ted Woodward, the Chairman of the Department of Internal Medicine, and Dick Hornick, the Chief of Infectious Diseases, gave me permission to work at the National Institutes of Health (NIH) for almost a year. In the fall of 1968, I joined the laboratory of Dr. Robert Chanock, the Laboratory Director of the National Institute of Allergy and Infectious Diseases on the Bethesda campus.

Chanock had an illustrious scientific career in respiratory infections, and was a brilliant investigator keenly interested in local secretory antibody.

In contrast to antibody that circulates in the blood like attack units in a strategic air command, local antibody is secreted onto the surface lining the mucous membranes, ready to defend against viruses like influenza, stopping them in their tracks before they invade the vascular highways of the body. Once in the body, the secondary defenses, including circulating serum antibody, are activated to neutralize pathogens. In geographical terms, secretory antibody is at the front line, the first to see action in the biological conflict with respiratory pathogens.

Bob Chanock was a mentor who liked to outline the long term plan, and he patiently sketched his ideas for the next 10 years of my life, an itinerary designed for a successful academic investigator. Up until that time I thought I would become a community-based physician, similar to the only role models of my youth. More than anyone previously, Chanock perturbed my thinking and sparked an interest in academic medicine and research.

I listened carefully to Bob, whose national and international reputation was launched with his discovery of a new cause of bronchitis and pneumonia in military recruits, *Mycoplasma pneumoniae*. He had made this discovery at the Marine Corps basic training base at Parris Island, South Carolina. This is the fraternal base to Camp LeJeune, North Carolina, where graduates of basic training would spend additional months learning more advanced combat skills.

The discovery of Mycoplasma was very important because recruits in high numbers were falling ill with pneumonia of unknown cause. The Gram stain of expectorated sputum failed to show the presence of any pathogen, routine cultures for bacteria were negative, and penicillin did not alleviate the symptoms. Later, Chanock and others would describe the fact that Mycoplasma, small rod shaped organisms, have membranes like bacteria but no cell walls. Without cell walls, the Gram stain is not taken up. Additionally, special culture media with uniquely required ingredients for growth were required for isolating Mycoplasma in the laboratory. Furthermore, because penicillin interrupts the assembly of cell walls, it has no effect on the wall-free Mycoplasma. Instead, clinicians now prescribe drugs like erythromycin or tetracycline that work by blocking the synthesis of vitally important microbial proteins.

I couldn't fathom the level of excitement and pride that must accompany the discovery of a new cause of disease or a new syndrome of illness. Moreover, I was especially impressed that a world class scientist was so generous with his time and earnest in offering me his advice. Chanock assigned me to work with Dr. Cal Perkins who was studying a vaccine for rhinovirus, the cause of the common cold. Working with Perkins were young physicians of my age, including Harry Gallis, who eventually would become a member of the medical faculty of Duke University, and David Tucker, an aspiring ophthalmologist, who was the son of the great metropolitan opera tenor, Richard Tucker. Though contemporaries, both Harry and David would teach me important aspects about the biology of viruses and mycoplasma, and I was invited to the weekly journal club where a few articles in the scientific literature were reviewed critically. David was a confident young man, particularly witty, and didn't take himself seriously. He was the key source of laughter in the lab.

The pace at NIH was steady, but not frenetic. The days were well planned, and people were relaxed with a comfortable humor that characterizes the world of a scientific laboratory. There are many periods where waiting is important, the down time needed as investigators pause for a chemical reaction to be completed or a centrifuge to take 20 minutes or more spinning down important components of a mixture. Compared to clinical training with days of admitting patients without any schedule while managing urgent or emergent problems on little sleep, the time at NIH felt like a trip abroad, and I was learning.

To test a vaccine for rhinovirus, Perkins would randomly assign half of a group of volunteers to the new vaccine and half to a placebo vaccine. Three weeks later he would instill a well characterized common cold virus into the noses of all volunteers. A virus was chosen that had sufficient virulence to infect 90% of those given that dose, the ID_{90}. The three-week interval was chosen because that time represents the biological defense buildup period when both local secretory antibody and systemic bloodstream antibody would peak in titer after initial contact with the virus in the vaccine. The human volunteers in this experiment were prisoners incarcerated at the Lorton Prison on the outskirts of Washington D.C.

In the special room where we would add the virus to the prisoner's noses, there were 10 to 12 bunk beds, and about 20 volunteers, almost all of whom were black. Each seemed quite pleased to take part in the clinical trial. I had suspected that the boredom of daily prison life was a key motivation, but I also learned fragments of a type of prison Mafia that controlled the list of volunteers. We were told that the prisoners who wanted to take part in the study would have to pay the controlling inmate boss in cigarettes in order to apply.

After all of the prisoners had received the drops of virus suspended in liquid solution in the nose, they were instructed to lie quietly for 30 minutes to give optimal chance for infections of the epithelial lining cells of the nose and pharynx. Our team with Perkins, Tucker, and me just had to sit and wait. After about 15 minutes, I noticed an old acoustic guitar in the room, and I asked the inmate nearby for permission to play it. I had learned various chords and strumming techniques at Haverford from my college roommate. The songs we had played for hours in the college dormitory were often inspired by antiwar protestors like Bob Dylan, Peter Paul and Mary, Joan Baez, and others.

I started playing and singing, "We Shall Overcome," not cognizant of the reality that in that room were three white physicians from NIH and almost 20 black prisoners, our volunteers lying on their backs in their bunk beds. By the second chorus almost all the inmates were singing along with increasing enthusiasm:

"We are not alone
We are not alone
Oh deep in my heart I do believe
We shall overcome some day."

By the third chorus, the volunteers were so animated that they began to sit up and shout out the song, the rhinovirus solution threatening to drip down onto their upper lips.

Perkins and Tucker were in a panic. If the prison volunteers didn't lie down immediately, the experiment and weeks of work might be ruined. I stopped abruptly while all became supine again and then finished the song. The group of prisoners had become exuberant, perhaps because the words had briefly removed some barriers between disparate social and cultural groups, had transiently connected establishment and disenfranchised, blacks and whites, and free people and detainees at a time of significant racial and social tension in the United States.

The emotion of that brief experience taught me to put a face on human volunteers of medical studies. They were not mere "study subjects" but individuals taking some risk and deserving of care and emotional support, even if prisoners of our society. It would be several more years before U.S. society defined vulnerable populations and proscribed research on such groups, offering safeguards to protect all volunteers including children, nursing home populations, retarded people and prisoners.

The rhinovirus did cause infection, but a series of subsequent studies showed that the vaccine was a failure. It did not protect against the infection. During that year, however, I learned about viral growth, mechanisms of infection, the proper way to do a clinical trial, and the method by which experts wrote up their scientific studies. I was introduced both to the world of laboratory science and a world behind guarded walls. With a great sense of pride I would also see my name on my first scientific publication along with those of my colleagues on the team and our mentor.

The formal period of my fellowship training in infectious diseases was coming to an end, and I knew that I had two years of duty in military service, a contract that I had signed a few years earlier.

My orders were to report for duty to the Marine Corps base at Camp LeJeune, North Carolina, on July 2nd in full uniform. The Navy had recognized that I had worked with Bob Chanock, that he had discovered Mycoplasma as a cause of important respiratory infections, and that I could work at a virus and Mycoplasma laboratory, the Field Research Laboratory at Camp LeJeune. The small lab was developed to monitor and study respiratory infections in military personnel at the base home for 30,000 recruits as well as the experienced troops who had returned from duty in Viet Nam. There would be only two physicians, but we would be assisted by 10 Corpsmen working as laboratory techs.

I left my wife and six month old daughter at home and proceeded to travel alone to Camp LeJeune because base housing was not yet available for us. There I was given a comfortable room in the BOQ, the bachelors officers quarters, and later I would seek temporary off base housing to rent. I read the book on protocol the night before duty to prepare my uniform for official wear, and I arrived the next day at 0-800—8 a.m. – to meet my new boss, Navy Captain Walt Beam. By 10-00, word had already come from the base commander that we were to be dismissed at noon and that we did not have to return to duty until July 6th. This directive astonished me, and the whole experience seemed an incredible waste of effort for me (having traveled from Baltimore for four hours of duty, and now languishing for three and a half days without my family.) The full time military personnel who were already settled on base with their families had an entirely different perspective, however.

At restaurants and cafeterias on base one could order grits at most meals, and piquant chili dogs and chili burgers at lunch. For dinner there were always fresh seafood options and delicious steaks. After dinner at the officer's club I entered the bar hoping to meet some of the other officers from the base. The bartender had designed a special drink that night in honor of the July 4th celebration. It was dubbed a "Cherry Bomb," a large tumbler with half vodka and half cherry liqueur. Only ten cents! After a few sips I quickly estimated that the lethal dose that would kill 50% of us inchoate drinkers, the LD_{50}, was about 60 cents worth. Although there were significant numbers of men taking advantage of the Independence Day special, I switched to beer and survived the holidays.

Walt Beam was a seasoned veteran of the Navy, a microbiologist by training, and a confident man with special talents for leading physicians. Walt too introduced us to the special food of the low country: steamed shrimp, raw oysters, fried flounder, and hush puppies -- the deep fried bread imbedded with spices. He was close to retirement and gave me and the other new officer, a pediatrician named Dave McCormick, very much a free hand. Whatever we wanted to study was fine, just keep him fully informed and permit him to review our protocols. With unusual common sense, he had broad experiences in various microbiological techniques. Walt was very supportive and carefully and willingly guided our initial efforts at scientific writing from naive drafts to the final copy to be submitted to major journals. This process might take nine or ten revisions or more before our papers were considered acceptable. There were times when I felt that he was drawing a moustache on my Mona Lisa. But I realized that I was learning to write science more clearly and succinctly and I was surprised, never having expected to find a senior friend and mentor in the military.

In late fall there were successive waves of adenoviral respiratory infection on base, each caused by a different family member. Cases of type 3 were followed by type 4, and then an explosion of type 7 occurred. The lessons in population morbidity and opportunity costs of respiratory infections were writ large in this closed community. My two years of military medicine convinced me that preventive strategies were quite appropriate, even essential if people are to feel well and be productive.

Winter at Camp LeJeune, North Carolina could be mean, especially in January when the temperature was just above freezing. A cold rain seemed to chill the body and rattle the underlying bones. On one of those bleak days, in my role as a new Navy Lieutenant Commander in the Medical Corps, I was asked to review some data with one of the Marine Corps colonels on the rates of respiratory infection among trainees. An officer is not issued an umbrella, and with a sudden and unexpected downpour, my uniform and I were soaked as I entered the Colonel's office.

Greeting me was a master sergeant with 20 years of experience, the stripes on the sleeves of his uniform indicating his extended tenure in the military. He was short, muscular, and very stocky with what must have been an 18-inch neck flanked by the bulging pillars of his sternocleidomastoid muscles traversing obliquely from his collarbone to the bony prominences below his ears. He had a huge grin full of teeth, filling most of his lower face. "Good morning, Sir! How are you today?" he inquired enthusiastically in his somewhat raspy voice.

"I'm uncomfortable Sergeant," I responded, plaintively. "With the icy rain I am shivering with this miserable weather."

He grinned again, all of his teeth telegraphing enormous optimism, "Yes, sir. It's another great day in the Corps!"

Caught off guard I felt immediately embarrassed about my whining over a trivial climatological matter in front of this experienced warrior who had just outlined a new perspective for me.

"Of course, it could be a lot worse," I offered apologetically. He was full of energy, focus of purpose, and a positive attitude. There could be no doubt that his trainees were successful.

I've always remembered the statement, and whenever I begin feeling sorry for myself, I think of that Marine Corps sergeant and I hear his confident voice once again altering my center of gravity and restoring emotional equilibrium.

Most of the epidemics of infection at LeJeune involved the newly graduated recruits from Parris Island, South Carolina, who were housed in a high-ceilinged, spartan room with 20 bunk beds along each wall, 80 people in the same platoon. Before being deployed around the world for permanent duty, they spent ten weeks gaining advanced military skills at LeJeune. The Marine Corps prepared their men for the challenges ahead. The Drill Instructors were seasoned veterans, tough task masters, and dedicated mentors. They stressed the concept that excellence meant doing it right the first time. At LeJeune the specialty training was intense – rehearsed until it was second nature, until the executions were performed with precision and success.

During their advanced training, recruits would often acquire a bronchitis or sometimes even pneumonia with Mycoplasma. Mycoplasma is an unusual respiratory pathogen that can cause relentless coughing, which in turn transmits this infection to those within a few feet of the large droplets expelled in fits of severe coughing. Sometimes there is an impressive pain in the front of the neck perceived as a "sore throat," but in fact caused by the inflammation of epithelial tissue lining the main windpipe to the lungs, the trachea. Many patients also complain of a dull frontal headache and low grade fever, and despite the pneumonia, the amount of disease noted on chest x-ray can be much less than the symptoms would suggest. In fact, the organisms don't invade the small air sacks of the lung, the alveoli, like most bacteria causing pneumonia, but instead attach themselves by specialized hooks to receptors on the tiny airways leading directly to the alveoli. Of interest, the infection causes an inflammation in the lungs that is triggered by the body's response to the microbes hanging out in the airways. In attacking our own cells, our immune system could be accused of overreacting to a feigned microbial advance on our lung tissue. In the process we are wounded by friendly fire.

With 80 young and non-immune recruits from all over the United States housed tightly together, it is not too surprising that an infection in one would quickly march through the other troops in the same platoon. An insightful scientific publication from that time reported that an infection in a drill instructor led to an outbreak among many members in a platoon. Those recruits unfortunate enough to have had their bunks located near the drill instructor's room were most at risk. Bellowing out orders raucously for extended time periods to the rigidly attentive recruits, a drill instructor could create repeated explosions of infectious droplets encasing the mycoplasma. The droplets could circulate around the barracks for short distances like microscopic hot air balloons about to be inhaled by the recruits. Two to three weeks later, an uncharacteristically long incubation period for a respiratory pathogen, the recruits would be at Sick Bay with symptoms, and some would be hospitalized, interrupting their training schedule and perhaps depriving them of graduating with their original cohort of trainees.

Like clockwork, the Marines could count on an early autumn epidemic of *Mycoplasma pneumoniae* bronchitis and pneumonia. In late fall there would be thousands of cases of adenovirus infection, and sporadically from fall to spring there would be 10 cases of severe meningococcal infection each year. As new physicians in the field research lab at Camp LeJeune, my pediatrician partner and I would record all of this respiratory misery for the military.

The lab with 10 expert corpsmen at the bench would identify all of the causes of acute respiratory infection among hospitalized patients. These microbiology technicians worked diligently, allowing my partner Dave McCormick and me the time to describe the seasonal trends, symptoms and signs, antibody responses and crude measures of impact to the military. I did not anticipate that the work of those two years would yield 25 publications, an astonishing record and one that I can attribute to the energy and fellowship in that lab.

Having fulfilled my military obligation, I was recruited as an Assistant Professor to the University of Virginia in the fall of 1972 to work on respiratory pathogens. I had not realized it at the time, but I would soon return to study Mycoplasma infections in Marine Corps trainees in boot camp. My new boss in Charlottesville, Jack Gwaltney, and I would receive a contract from NIH to study a *Mycoplasma pneumoniae* vaccine in Marine Corps recruits at Parris Island located just outside the city of Beaufort, South Carolina. At that time I had just begun to train specialists in infectious diseases: physicians who had completed a specialty in Internal Medicine could enter what is called a "fellowship" to train in a subspecialty for two to four more years.

My secretary mistakenly booked me and my first fellow in training, Bob Craven, in Beaufort, North Carolina, a city near her hometown. When Bob and I arrived at the Holiday Inn in Beaufort, South Carolina, we learned that a local regatta had attracted thousands of visitors that weekend, and no rooms were available in the city. However, we were told of Fripp Island another 20 miles into the ocean, and if we agreed they could reserve a room for us in the only hotel there. When we arrived to find a comfortable hotel located directly on a wide beach, we could barely contain our elation, wryly uttering to each other our new slogan for the vaccine project, "War is Hell!"

The Barrier Islands of South Carolina had almost deserted white, pristine beaches, and the rhythmic sounds of the waves were reminders of comfortable summers spent as a child on the coast of New Jersey. We toasted our good fortunes over dinner that night and reviewed our schedule for the next day. Parris Island was a 45-minute drive across a series of small islands and tiny villages with names like Frogmore, Lady's Island, and Hunting Island. Bob Craven and I would assemble a local team to support the trial of the new Mycoplasma vaccine: a retired Marine Corps Gunnery Sergeant, a retired Navy Corpsman, and a retired Army General all accepted our advertisements. Later I would ascribe the success of the study to our having selected a "few good men."

Over 8000 Marine Corps recruits volunteered for the study. Half would receive vaccine and half placebo in a properly-conducted, double blind study. Neither the volunteers nor the investigating team members knew what was given until the code was broken at the end of the study. Blood was drawn from all volunteers at the beginning and end to see which men developed new antibody to Mycoplasma, evidence of an immune response to interim infection. All cases of new pneumonia were cultured, and new sets of blood drawn for precise timing of antibody response.

When recruits are first assembled to form a new platoon, they are given their platoon number and a unique laundry number for identification. For example, platoon 180, laundry number 11 always identifies Private John Q. Smith. He would keep the same identifying numbers even if he graduated after his cohort. The men with newly shaved heads stand at attention with only boxer shorts on, not moving a muscle. A designated recruit prints the laundry number, 1 through 80 in succession, onto the bare chests of the recruits with an indelible red felt pen. When we learned this, we investigators found a sure method to track all volunteers, equal in specificity to an individual's social security number or a fingerprint.

The young men received the vaccine via a "gun" that shot the liquid below the skin surface to the subcutaneous fat and muscle for immune processing by white cells called lymphocytes. The platoon's drill instructor paced rhythmically back and forth, Smoky-the-Bear style hat in place, barking an occasional order. Recruits never talk unless asked a question by a superior, and they never say the word "I" in response. Only the 3rd person is tolerated: "The private thinks…, Sir!"

When we walked up and down the line of troops just vaccinated, I recall asking the first volunteer if the vaccine hurt. "No, sir!" he quickly responded. The second recruit also loudly affirmed, "No, sir!" The third recruit seeing the drill instructor in the corner of his eye responded even more enthusiastically, "Felt good, Sir!" The vaccine sometimes gave a red area on the skin 2 to 3 inches in diameter, yet our reported side effects profile would clearly be understated.

Young men who had never left home and whose contact with the world at large was limited were among those who volunteered enthusiastically for the Corps. One day we observed the drill instructor giving the visual acuity exam to a recruit. Standing 20 feet from the eye chart, the drill instructor held a pack of Camel cigarettes over one eye and told the recruit to read the chart. There was no response. The drill sergeant came closer to the timid recruit and reissued the command, "Read it, Private!" Still no response and now the recruit seemed genuinely puzzled. The drill sergeant moved ever closer, with the wide brim of his hat now touching the forehead of the recruit, as he yelled in his guttural voice, "I said, read the fucking chart, Private!" At this point the recruit promptly responded, " Sir, the private can see the chart, but he can't pronounce those words!"

There were also hints of the zealousness of the recruiting process on the parts of both recruiters and recruits. I saw a recruit in formation with a linear elevation of skin traversing the side of his recently shaved scalp, and he admitted to me that he had had a shunt in his central nervous system, bypassing an obstruction of the flow of cerebrospinal fluid from the large lakes of the ventricles in the brain to the fluid surrounding the spinal cord. I was viewing the outlines of a plastic catheter just below the scalp that had been previously inserted by a neurosurgeon to maintain the normal fluid pressure in the brain. Another recruit who was limping painfully after a 20-mile march was found on x-ray to have a prosthetic hip. I assumed that both would be dismissed.

Those who managed to stay in the platoon for the entire training period were in great physical and mental shape, were Gung Ho Marines, and appropriately proud of their accomplishments. They would go on to graduate and be assigned to advanced training at Camp LeJeune. Like organisms exposed to a hostile environment, those who survived the ordeal were selected because of their extraordinary abilities to adapt. They would succeed.

A few weeks after the launch of the study, a group of three senior scientists from NIH came to visit our team and observe the conduct of the clinical trial. All had PhD's, and none were physicians with clinical experience. They too stayed at Fripp Island, and we left for Parris Island at "0 dark 30," about 5:45 a.m. By 8 a.m. it was 85° Fahrenheit and by 9 a.m. 90° with 90 -95% relative humidity. The NIH trio of monitors was fatigued by the weather, and we told them on a day like this to expect heat related illnesses.

Before 10 a.m. that day, two recruits developed heat stroke, their body temperatures rising rapidly to 105° – 106°F. The Corps was aware of the risk, observant, and well prepared. There were two oversized bathtubs filled with ice and water in anticipation. The victims were placed in the freezing water, clothes removed, a rectal thermometer placed, and their limbs rubbed vigorously until core temperatures fell to a safe range, about 101°F. At that time it was reasonable to transfer them to the hospital. By standards of any medical care I had witnessed, this was barbaric treatment. But it was rapid, effective, and at least in young men, everyone survived.

A confident drill instructor stood over his recruit with a certain paternal reassurance and remarked positively, "Don't worry, private, we'll have you back on the field by tomorrow!" The recruit was shivering and frightened, and for a brief period I could only imagine whether he was more fearful of survival and a return to the same fate the next day, or of his currently morbid condition. My guess, however, was that he would be eager to return to his own platoon as soon as possible. He had a deep commitment to the Corps.

After lunch the inspecting team from NIH who had witnessed our vaccine program, the two cases of heat stroke, and the extreme heat and humidity now looked whipped. They commended our efforts, wondered out loud how we managed to adapt to the elements, and suggested that an early afternoon return to Fripp Island was appropriate to cool off by the ocean. A few weeks later their final site visit report on our efforts arrived. It was quite laudatory.

The Mycoplasma vaccine was 70% protective, and no side effects occurred except the mild to sometimes moderate redness of the skin at the site of vaccination. Seventy-percent reduction in cases of pneumonia is very good but not outstanding for vaccines targeting only a special population, especially for an organism causing mild fever and rarely death. Yet we felt that we were part of a successful and useful clinical trial that had shown efficacy. I don't know what problems may have been encountered in attempts at producing large batches of vaccine or of the cost benefit studies of giving it to recruits. It is likely that at that time a Mycoplasma vaccine would have had indications only for military recruits. The vaccine was never licensed, never deployed. Looking back on these experiences, however, I have to admit it was another great day in the Corps.

Mycoplasma, a specialized bacterial pathogen, introduced me to the military and to epidemics of respiratory infections in young men in confined spaces. It has its preferred season, and we still don't know why. What pressures induced this bacterium to evolve without a cell wall remains mysterious even today. Mycoplasma also introduced me to the world of vaccine testing and properly conducted clinical trials. Having worked for several years in its testing, I was impressed with the vaccine's efficacy and emotionally committed to its acceptance. It wasn't at all clear in the 1970s how prevalent this organism was in general. However, in the year 2005, *Mycoplasma pneumoniae* is recognized to be the cause of 10% to 20% of all community acquired pneumonias among civilians, mostly in young adults.

With my experience in the Navy and my collaborations at both Camp LeJeune and Parris Island, I gained enormous respect for the military and especially the Marine Corps. Like the body's sophisticated immune system they are programmed to respond to the most dangerous of assaults, committed to defense regardless of the hazards, and always ready.

When I am asked by young faculty members what I think it takes to prepare to be successful at research, I respond that there are many elements, but I begin with the following: excellent training, a gifted mentor, a quest for excellence, a positive attitude, energy, focus, persistence, experience, and an ability to adapt to change. I would add that research is a process, not an event. And there are few lone rangers; it requires a team to be successful, but a team led by an experienced person capable of clear and independent thinking.

I was fortunate in the late 1960's that during my specialty and subspecialty training to have witnessed much of what I am speaking about. I was not consciously aware of each aspect but became more cognizant of the value of each as I witnessed a parallel with my subsequent experience in the military.

There were other important lessons, those dealing with the complex relationships of an individual to the group. A person's identity can be imposed on them by the group as I saw both in prisoners and Marine Corps recruits. In some cases such as the military, belonging to the group is absolutely essential, and the loss of some individual identity seems worth the cost of preserving the safety of a country.

When I was training it was considered ethical to conduct experiments on isolated groups to take advantage of individuals whose grouping was imposed by society. They included retarded children, prisoners, and military recruits. Many have agreed that such human experimentation caused no harm, benefited society, added utility for such people or at least that "more good than harm" resulted.

There is some truth to all of this, but some "volunteers" became injured in the experiments. More importantly, they were not fully "free" to choose to participate or not, could not be fully informed of the risks or benefits. Did the black prisoners really understand the study design of the common cold experiment? Would they have been able to differentiate an experiment in which the worst that could happen would be a runny nose from one in which a serious pneumonia might occur? Would they have volunteered had they not been part of the large prison group? Would the young Marine Corps recruits have been volunteers without their identities intertwined with those of their respective platoons? Did they perceive their participation as necessary, as patriotic?

In my preparation for an academic career, I had thought that prison and military volunteers were fair subjects of experiments so long as the side effects were minor. But I also knew of experiments in which the subjects were given doses of more serious infections including Tularemia (rabbit fever), Typhoid fever, Rocky Mountain Spotted fever, viral hepatitis, and others. Some would become quite ill. Recent experiences among volunteers being studied at prominent medical universities confirm that deaths can occur.

To me, human volunteers need the same reverence as patients. Neither would choose to be the subjects of our attention as physicians. They are guests in our medical institutions. Furthermore, in addition to whatever identity that they have as individuals and as members of a group, that identity is further perturbed by their being in the hospital, by being the subject of an experiment and identified by some number, today usually their social security number.

In the 30 years since I left my training, there have been many safeguards in place to protect volunteers. Prisoners and retarded children are no longer acceptable study subjects. Experiments on military recruits are quite limited. Most study subjects come from the community, and informed consent is a serious process in which the risks and benefits are explained carefully, where the volunteer is told that he or she can stop at any time without consequence. Yet the newspapers are replete with experiments gone wrong. Usually the problem is the lack of focus or carelessness of the principal investigator.

Are the safeguards sustainable? In the mid 1970s, I once spoke with Dr. Saul Knugman, the famous pediatrician who helped identify the fact that Hepatitis A and B were different viral infections of the liver. In a series of experiments, he did this by injecting serum contaminated with both viruses into retarded children. Because the incubation period for hepatitis A virus is short (2-8 weeks) and for hepatitis B virus is long (2-8 months), these children developed waves of two infections of the liver. Dr. Krugman was undergoing severe criticism at that time for his studies, not because any child had died and not because the science was suspect. His studies were elegant. But with the uncertainty of infecting doses, there was considerable risk of giving a huge inoculum. Most importantly, retarded children are not free to consent.

Dr. Krugman told me that he began his studies during World War II and that experiments on military volunteers were being conducted by all major combatants – the U.S., Germany and Japan – because this infection of the liver, viral hepatitis, was of such strategic importance. Even during the subsequent Korean War such experiments continued. However, with the end of the Viet Nam conflict, the rights of the individual became more important, and Krugman's research came under fire.

There may be a general lesson here. In times of war, the public decrees that the rights of individuals decline and the importance of the group becomes paramount. As conflict recedes, the rights of the individual become extremely important. Krugman failed to have his fingers on the pulse of society and kept his experiments going even after the winds of war had died down.

As I ended my period of academic training and began my professional career, these were some of my thoughts. I would subsequently conduct several large studies with human volunteers: vaccine studies to prevent various illnesses, topical nasal creams to prevent serious *Staph aureus* infections, and several studies to treat life-threatening sepsis and bloodstream infections with novel therapies. My training and subsequent experience have given me a perspective of the value of the individual, the complex relationships of individuals to groups, and the need continually to be vigilant about preserving individual rights.

As I now entered an academic career, some further details of the qualities of a physician continued to become apparent, the distant silhouette beginning to take on distinctive features. Despite an exciting training period, however, I realized that my experience in managing individual patients and infections in populations was limited. I also knew that I might meet challenges which were not clearly outlined in textbooks. At that time I had assumed that if questions arose, senior faculty members at the

Medical School would know the "answers." I would soon learn that this assumption was wrong, but that would be my good fortune.

PART II

PROFESSIONAL LIFE

DISCOVERY
(Rickettsia)

In the early 1980s, a totally new and deadly illness visited itself upon a small number of citizens in the Charlottesville, Virginia area. In the span of one month, five seriously ill people were infected with an organism causing an illness never previously described. During the subsequent 20 years of my career I can say that the intensity of my focus on what was happening to those patients has never been exceeded. The diagnosis rested on repeated questioning of patients whenever possible, otherwise inquiring of their family members and friends. We attempted to trace their every move for the 2-3 week period prior to their illnesses. Their common exposures were teased out of a myriad of facts, and combined with similarities in observed findings on physical examination, shared pathological and serological features, they formed the architecture of a new syndrome. That process highlighted the best of scientific collaborations.

The faculty in the Infectious Diseases division at the University of Virginia at that time would spend three months on the inpatient services each year, two while leading the infectious disease consultations and one directing an internal medicine ward. Our clinical team usually included a senior resident, two or three interns, and three medical students. A few of us who particularly enjoyed managing acute illnesses chose to spend a month in the medical ICU. In critical care units there was a limited number of patients, an intense experience with a close-knit critical care team with one-to-one nurse to patient care, a high level of excitement, and an opportunity to see complicated, acute infectious diseases. With a limited number of patients we also had special opportunities to become more familiar with them and their worried families than was usual with a non ICU ward assignment. In back-to-back assignments on the infectious diseases services and the medical ICU, an unusual cluster of patients with a new syndrome of illness came to our team's attention.

A 41 year old woman who was a mother of five and a nursing student had been seen in the emergency room, and after an evaluation was discharged with a diagnosis of a viral syndrome, two days before I met her. Subsequently, on admission to the hospital she complained of fever, severe headache, and remarkable difficulty holding her head up, nausea, and pain in her eyes when exposed to bright light (photophobia). Whenever we see patients with unexplained fever, infectious diseases specialists ask about medications, exposures to ill contacts, travel history, insect bites, and exposure to animals.

Seven days before she was admitted to our service, the patient had butchered a deer that had been skinned by her husband. In the weeks before her illness, she frequently walked in the woods with her flea-infested dog.

When we first saw her, she was drowsy but had no evidence of a stroke. She was oriented: she knew who she was, where she was, and could identify the correct day and time. Within two hours, however, she suddenly developed an unusual form of seizure beginning with a twitching of her left hand and turning her head to the left. Quickly thereafter she developed signs of brain swelling and downward herniation of the brain through the narrow opening of the bony skull with resulting pressure on the basal areas, the anatomical sites of cranial nerves controlling eye motion and breathing.

An emergency cat scan (CT) of her head confirmed massive edema of the brain, the swelling resulting from diffuse inflammation. Furthermore, on the first hospital day her heart began to beat erratically, an ominous sign in a deteriorating patient. Despite receiving several classes of antibiotics intravenously, she died 26 hours after admission.

The overwhelming sense of the medical team was one of frustration and disbelief that a patient could progress to death so quickly despite heroic efforts. There was also a feeling of defeat, a let-down of mood and for many a sense of guilt along with the sadness. At the time we did not recognize that we were witnessing a new syndrome, and when disease progresses in such an unannounced and shocking way we wondered if we did not anticipate immediately the gravity of the situation.

If this was an infection, what could be its cause? Aside from meningococcal disease, none of us had ever seen an infection advance so relentlessly to death in so brief a time. Even unusual forms of viral encephalitis advanced more slowly. A massive stroke could do this, but the fever and multiple areas of the brain involved made this unlikely. Perhaps a poison, but no tests were positive. In fact, no one on the team could even think of poisons causing the devastating constellation of signs and symptoms we witnessed.

At autopsy her brain was severely swollen with marked congestion of all blood vessels. There were hemorrhages in various sections of the brain. To the surprise of everyone there was inflammation of the blood vessels themselves, with the white blood cells seen infiltrating the walls of the small veins and capillaries especially. The mother of five children had died of an acute form of a vasculitis, and our medical team had nothing intelligent or informative medically to console the family. We began by telling the truth: we had never seen anyone progress so quickly and under our eyes to a rapid death. We suspected an infection of the brain, possibly a viral encephalitis, but it was impossible to say which virus could do this. We asked permission and received it for a post mortem examination so that we might find answers for the family. We offered our sympathy but admitted that we couldn't begin to appreciate the enormous pain they were experiencing.

We began to review the medical literature for reported cases of acute arteritis of the brain. Rarely viruses could do this, such as *Herpes simplex*, but this pathogen involved large not small vessels. Cytomegalovirus was occasionally reported to do this in individual cases. Rocky Mountain Spotted Fever could do this, but it usually involves small vessels throughout the body, not just in the brain. There had been two isolated reports of *Mycoplasma pneumoniae* associated cerebrovasculitis, but tests in our patients would later rule out Mycoplasma.

Four days after the tragic death of this unfortunate woman, a 59-year-old man was transferred to our service from an outside hospital with fever, confusion, and fatigue. He had had recurrent seizures and required tracheal intubation for airway control. According to family members he had owned a dog that was heavily infested with fleas. An examination of the retina of his eyes with an ophthalmoscope revealed that he had small occlusions of the tiny retinal veins and arteries consistent with an inflammation of the vessels. His eyes preferentially gazed to the left, toward an inflamed area of the brain. He had a stiff neck, evidence of meningitis, and signs of serious neurological defects involving many disparate parts of the brain. Despite full critical care support and receipt of multiple antibiotics intravenously, he remained comatose and died several weeks after admission.

When doctors have two quick and unexpected deaths among their patients, we begin to wonder what dreams of theirs went unfulfilled, what hopes for the future never reconsidered, and what deep feelings were never expressed – perhaps put on hold – by a person unaware that death was so near.

A brain biopsy specimen showed similar findings to the first patient, inflammation of the small vessels of the brain, a vasculitis. Could this man have had the same disease as our first patient? Was this a transmissible disease? The two patients had not known each other, had never met. Importantly, none on our team had ever seen a patient with fever and acute vasculitis of the brain, and now we had two and two sudden deaths. None of the senior physicians who had lived for decades in Virginia could recall seeing any disease similar to those of our patients. They had no answers to this mystery.

My generation of physicians was weaned on the concept of survival after disease. In the early 1980s, we never considered the cost of medical care or cost-effectiveness estimates and only occasionally addressed quality of life issues. Our jobs focused on life and death. When we framed ethical questions we gave great weight to the concept of beneficence – do everything possible for the good of the patient. We did everything to prolong life, perhaps turning a blind eye even in the face of terminal illnesses. With acute disease, we went full throttle and did everything we could to diagnose and treat the problem. Any considerations of balanced value for the society were unheard of. Such ethos was so ingrained that the loss of patients to a devastating infection, especially one that was not understood, was painful. I felt helpless in the face of this syndrome and had a nagging worry that perhaps these were unusual presentations of known diseases. Perhaps there was something I might have done to reverse the fate of these two patients, real people with caring families now experiencing terrible and unforgettable grief.

Only two days after we had admitted our second patient, a 41-year-old man arrived with fever, malaise, headache, and both nausea and vomiting. In the ICU, he was confused, combative and was seeing double, the result of sudden loss of coordinated vision from disease of one of the cranial nerves controlling eye muscle movement. Remarkably, he too had been in the woods and had cut wood in the 10 days before his illness. He also had a dog infested with fleas.

He was immediately begun on several antibiotics including chloramphenicol, a drug with broad spectrum activity against a large cross section of unusual bacterial organisms. Even in the 1980s, we reluctantly used chloramphenical because it was known to suppress activity in the bone marrow, the site of manufacture and processing to maturity of red cells, white cells and the platelets, the clotting cells of the body. Furthermore, rarely it could cause a complete shut down of the bone marrow, the fatal aplastic anemia syndrome. But we needed to try something new and opted to add chloramphenical.

Repeated CT exams also showed brain edema, the swelling that invariably accompanies inflammation. He developed seizures that even after his discharge from the hospital were difficult to control. Clinically, the involvement of multiple areas of brain suggested a vasculitis. We now had three patients.

We had saved sera of all patients from the days of admission, and when possible, follow-up specimens were obtained three weeks later. If a patient was subsequently identified to be infected with an unusual organism, his serum might show evidence of a rise in specific antibody titer to the pathogen from the first to the second of the paired specimens. Examining serum for the new development of antibodies is akin to finding footprints in the snow: although the organism is gone, we can observe reasonable details indicating that it made its way here recently.

The fourth patient, admitted on the day after the third, was a 25-year-old truck driver with new onset seizures. For three days he had had fever, malaise, severe headache, muscle aches, nausea, vomiting and dizziness. On questioning he responded that he had a flea infested dog and had handled firewood in the 2 weeks before his illness. Was the exposure to wood a sign that some animal living in the wood harbored the pathogen infecting and killing our patients? Or was it a surrogate marker for being outdoors, and some organism living on a wild animal might have become an accidental tourist crossing species lines to infect our patients? Could fleas be transmitting a serious infectious agent? These were questions we began to ask.

The young man too had some stiffness of the neck, but his CT scan was normal. Nevertheless, he was begun on the antibiotic, chloramphenicol. A brain biopsy confirmed acute vasculitis of the small arteries, both the capillaries and venules. He required intubation for almost 3 weeks, but was quickly weaned from the respirator, and at that time was discharged well after a 30-day stay.

About two weeks after the fourth patient arrived, a 16-year-old high school student was admitted to the hospital with fever, headache, double vision and incoordination. The latter symptom was the result of disease of the cerebellum, the body's coordination center situated at the base of the brain near the highest part of the spinal cord. He also had slurred speech from the incoordination of the muscles orchestrating phonetics, cranial nerve dysfunction affecting eye muscle movement, and other evidence of severe central nervous system disease. In the 2 weeks before admission, he had been hunting in the woods and had killed, skinned and eaten two squirrels. Having received intravenous chloramphenicol, he was discharged after two weeks, but at that time had residual slurring of speech and was walking with some lack of coordination.

We knew that we had an epidemic of a disease that none of us or our many colleagues had previously experienced or could explain. All five patients had clinical or pathological evidence of vasculitis, inflammation of the vessels of the brain. They all had had fever, and we named the syndrome simply, "Acute Febrile Cerebrovasculitis." All had lived within 60 miles of Charlottesville, and none had traveled outside of the state of Virginia. All had been in the woods and had dogs infested with fleas, but none had recognized any exposure to ticks.

All had had spinal fluid examinations, but the Gram stains and cultures were negative as were cultures of the blood. The full name of the discoverer of this simple stain was Hans Christian Joachim Gram, born in Copenhagen in 1853. Throughout his life his passion was botany, a science in which careful classification of species was required and the use of the microscope was common. The study of plants was also the basis of his interest in pharmacology and may have stimulated his interest in pursuing a medical degree, which he received in 1878. The 1880s were explosive years in Medicine and Science. Pasteur confirmed his germ theory of disease with vaccine prevention, Koch discovered the cause of tuberculosis, Ogston gave the name Staphylococci to certain pus-forming organisms, and Gram devised an enhanced staining process to differentiate causes of infection.

Gram particularly wanted to identify the true cause of acute pneumonia. He had been aware of the work on staining bacteria introduced by Paul Erlich with a dye called gentian violet. Erlich himself was also especially famous for his dogged attempts at finding a cure for syphilis, and after 606 tries he found his "magic bullet," an arsenical that was the first effective treatment. Gram did not design a stain to find bacteria in sputum or blood cultures, and he failed to appreciate the value of the stain. Nor did his colleagues understand its utility initially. In the 1880's, the common causes of pneumonia were not known, and the science at the time suggested that staining the lung tissue of patients who had died from pneumonia would be state-of-the-art science. The technical problem was that existing stains could not differentiate the lung cell nuclei from bacteria. What was needed was a staining system that decolorized the nuclei of the cells that were infected with bacteria and that would give a striking color to the microbial pathogen. The stain of Gram did just that. It also helped resolve a bitter polemic of the 1880s on the cause of bacterial pneumonia.

Gram was working in the laboratory directed by noted microbiologist, Carl Friedlander. At the time two different organisms were seen to cause pneumonia, the one that stained blue with Gram's method and was the most frequently observed (we now know this was pneumococcus), and the other that lost its initial blue stain was often rod shaped, and seen less frequently as the cause of fatal pneumonia (we now know this is the Gram negative *Klebsiella pneumoniae.*) At the time it was not appreciated that two distinct bacteria could be causes of the single clinical entity of pneumonia.

Different camps lined up behind each supporter, and it would take two or more years for the recognition that both groups were right, that more than one organism could cause a pneumonia. And still it would be a few more years before the true value of Gram's stain could be appreciated.

After 11 years of marriage, Gram lost his wife to pulmonary tuberculosis, an infection that was responsible for 25% of all deaths in Europe and the United States at that time. And after his retirement as physician in Chief at the Royal Frederick's Hospital, he returned to his passion of botany, hoping to find pharmaceuticals to prevent tuberculosis. Gram died in 1938, a few years before the announcement of Selman Waksman's discovery in the United States of the first drug for tuberculosis, streptomycin, not from plants but from chemicals elaborated from organisms living in soil.

In our own patients whose Gram stains were negative, extensive blood tests for antibodies to tick and mosquito - borne viruses that cause encephalitis were also negative. None had evidence of infection with any of the Herpes virus family including cytomegalovirus or EB virus, the agent which can also cause infectious mononucleosis. None had demonstrated serologic evidence of antibody development to unusual bacterial infections that are related to exposure to cattle or deer such as Brucella or Tularemia, respectively. All tests for syphilis which can cause a vasculitis, were also negative. More common viruses were also ruled out such as influenza and mumps, and some unusual causes of meningitis were ruled out such as lymphocytic choiomeningitis virus and many others.

The break in the cases occurred only after the reports of immune responses to a member of the Rickettsia family of pathogens. The four patients with paired serum specimens showed very modest rises in titers in the three weeks interval from admission serum to convalescent serum samples to *Rickettsia rickettsii*, the small coccobacillary organism causing Rocky Mountain Spotted Fever. Special studies performed by a friend and colleague at the New York State Health Department, however, found much higher titers or rises in titers to a sister organism causing flea-associated typhus, *Rickettsia typhi* but not *R. ricksettsii*. None, however, had high titers to *Rickettsia prowazekii,* a squirrel associated disease reported previously in Virginia. In summary, most tests supported infection with a typhus-like organism. Since the disease primarily and prominently involved small blood vessels of the brain, this syndrome was clearly distinct from both Rocky Mountain Spotted Fever and classical typhus, however.

Rickettsia are unusually small bacteria that must live within cells to be viable, and are not "free" to live outside of cells like most bacteria. They have pleomorphic shapes, sometimes appearing like rods and other times like cocci. The best recognized member of this family is called *Rickettsia rickettsii*, the cause of Rocky Mountain Spotted Fever, a tick borne illness that causes fever and rash. In Rocky Mountain Spotted Fever, the Rickettsia invade the endothelial cells lining the small capillaries and cause a vasculitis. If untreated, the inflammation leads to occlusion of the blood supply to vital organs, and the patient complains of fever, severe headache and especially muscle aches. The rash begins as a salmon-colored spot on the palms and soles and migrates centrally to involve the chest and abdomen. In the last stages of disease, small dot-like hemorrhages called petechiae appear, the patient becomes comatose, and a physician might have difficulty distinguishing this patient from one in the late stages of meningococcemia. As a rule, infections with Rickettsia cause a vasculitis that can be life threatening if not treated very early. Importantly, none of our patients had a history of exposure to ticks or lice, and by deduction, if we were seeing the effects of a Typhus-like rickettsia, the vector was most likely a flea.

My mentor from residency, Ted Woodward, was the first to have treated Typhus with chloramphenicol, an antibiotic that disrupts the assembly of amino acids that form key bacterial proteins. And our last three patients who received early doses of chloramphenicol survived.

A few years later in studies with colleagues in New York led by Dr. Karem Hechemy, we found further evidence suggesting a rickettsial infection by using very sophisticated laboratory tests called immunoblot studies. These are quite useful tests because they are specific for a particular protein or carbohydrate unique to a microorganism or a closely related family member. Currently, such tests are used to confirm infections with the Lyme bacterium and the Human Immunodeficiency Virus (HIV), the cause of AIDS. The immunoblot test told us that whatever caused the infections, the organism shared many biochemical features of a Typhus-like Rickettsial organism.

Our scientific reports stated the medical facts as best as we could describe them, and most newspapers around the country provided accurate coverage of our epidemic. Not every lay press article limited their reports to available science, however. At the time we were working up the epidemic and treating the patients, unusually colorful hypotheses were advanced in a report from a fringe press, the *National Examiner,* focusing on the epidemic with the riveting headline, **"Alien Virus Strikes Suburbia."** Describing some of the facts, the article proceeded to say that "menacing aliens from another planet have left a trail of terror behind them in the suburbs of a rural American town." Then there followed an alleged quote from a member of the Imperial Institute of Astronomy in Tokyo: "We tracked the course of an unidentified comet–like object that plunged to earth over the state of Virginia, and there's every possibility that it was an alien spacecraft." I was quoted as saying that there was "something special" about the infecting agent. Remarkably, a statement of an "officer from a local UFO society" noted that they had "discovered a circular burnt clearing 60 feet across in the densest part of the forest... I believe that we have a landing of a space craft, the occupants of which have somehow, purposely or not, infected the local wildlife with an alien strain of bacteria not previously seen on the planet earth."

There is some irony in the fact that in medical circles, it is considered to be very fortunate to be quoted in journals like the *National Examiner*. It is often a distorted representation of the science, and maybe that is the appeal for the lay readers. We all rush to make slides for our scientific presentations, knowing that by taking a perspective of such readers we are really making fun at ourselves, and audiences always like that at serious scientific sessions.

The version of discovery offered by the *National Enquirer* surely differs from reality. In truth, discovery is hard work requiring discipline, tenacity and open-mindedness. There is the absence of smoking guns and the need to re-examine what we know in fact, rather than to haul in convenient interplanetary explanations.

After publication of our manuscripts, I began to give a series of talks on this syndrome and offer the evidence we had for a rickettsial etiology. The organisms of this genus of bacteria are named in honor of Howard Taylor Ricketts, a creative investigator who identified the cause of Rocky Mountain Spotted Fever while working in a laboratory in the Bitterroot Valley of Western Montana. The tick had been recognized to be the carrier of the infection, but to find the true causative organism, Ricketts developed an experimental model of infection using guinea pigs.

In that part of Montana the mortality after spotted fever in the preantibiotic era of the early 1900s could be 80-100%. The local Nez Perce and Flat Head Indians who named the area after the bitterroot plant, *Lewisia rediviva*, believed that evil spirits caused the illness each year visiting the valley in the first warm days of Spring.

By 1909, Ricketts had solved the mystery by identifying the causative organism and asked the State Legislature for further support for the development of a preventive serum. However, when the legislature had not responded by the Spring, Ricketts decided to travel to Mexico City where an outbreak of Typhus was newly reported. It was in Mexico that Ricketts astutely recognized the clinical similarities of Rocky Mountain fever and Typhus, and a year later, in April 1910, he would announce that he had discovered the microorganism that caused Typhus.

With such enormous success behind him, Ricketts decided to forsake his globe trotting adventures and focus on an academic career. But at the end of April, however, when he was about to leave for a professorship and Chair of Pathology at the University of Pennsylvania, Ricketts himself contracted typhus. The infection was transmitted accidentally in the laboratory, and on May 3, 1910, Howard Taylor Ricketts had died.

In his quest for discovery, Ricketts would appear to have had insatiable curiosity and boundless energy. He ignored the Colorado State Legislature's failure to seize an opportunity and traveled over 1,000 miles to central Mexico for another opportunity. The nemesis of investigators working with deadly pathogens has throughout medical history been the risk of serious infection and death. In his search for novelty, Ricketts paid the ultimate price.

Our five patients in Charlottesville with a new disease, presumed to be caused by an organism named in honor of this creative and industrious investigator, had tragic consequences. It may be that the survival of the last three patients was related to their younger ages or possibly to their having received more doses of the antibiotic, chloramphenicol. Looking at the numbers – 5 unfortunate people among the 150,000 in the catchment area of UVA – one would like to know, why them? An epidemiologist asks, what specific risk factors such as flea contact, walking in the woods, or exposure to wild animals led to infection and illness? A lay person often asks what dietary, smoking or lifestyle indiscretions might have charted the course of the ill-fated disease.

In 1989, three years after our report, authors of a subsequent article reported a single case from another institution in which the patient developed serological antibody rises to *Rickettsia typhi*. This patient was from Ohio and showed similar symptoms and signs as our patients, and the authors stated that they had observed another case of "acute febrile cerebrovasculitis." Serum samples from a mouse trapped at the patient's home showed antibody to *R. canada*, a related species, raising the possibility that a rodent was the source of an unusual Rickettsial infection.

There are thousands of viruses and bacteria that are part of the permanent or transient flora of animals that have not been recognized or studied. Very occasionally events conspire so that these organisms jump species and infect man, by definition a "zoonosis" infecting people. Sometimes the infection remains endemic in people by virtue of person to person transmission like HIV. Sometimes there are repeated transmissions from animal to man via insect vectors such as occurs with Rocky Mountain Spotted Fever or Lyme disease, with organisms carried from animal hosts to people via ticks. And occasionally a set of circumstances creates a unique transmission that is uncommonly seen. I suspect that our cases represent the latter.

Current evidence supports a Rickettsial organism or at the very least, an organism that shares similar antigens, the protein or carbohydrate components that give very specific cross reacting antibody responses. Was the uncharacteristically warm fall of 1983 a factor in supporting the growth of an unusual pathogen, its animal host or its transmission? Was the usual peak in late fall and early winter of the flea population in animals in some way enhanced that year? I don't know what animal reservoir was most important or even if fleas were definitely the vector, but sometime in the future we will see this syndrome again, perhaps identifying the exact species and confirming the mode of transmission.

Ours was not a total triumph in medical discovery. The fact is that I never confirmed with certainty the cause of this disease. I did my best and came up short; a silver metal but not the gold. I was greatly satisfied, however, that I recognized and described a new disease.

I will always be grateful for the opportunity I had at UVA to recognize a new syndrome and work with an energized team to describe it. I learned that opportunity is an intermittent visitor, always arriving in disguise and unannounced. Like a shy guest at a large reception, if we fail to recognize and engage her, she moves quietly in another direction. If only we could pause briefly to admire her unique countenance, she would reward us with an illuminating romance. Discovery sparks the heart to beat excitedly, and the possibility that discovery could eventually lead to prevention or to new therapy is what gives additional vitality. Similar experiences in institutions across the country define the academic mission.

I had recognized my interests in finding unique information for some time. But the experience with these five patients fixed my resolve for clinical investigation in academic medicine. Academic medicine has offered me the gift of adventure, of seeing more unusual presentations of disease than are seen in the community, and of the time to pursue the clues to discovery. Many physicians practicing clinical medicine in the community share the same drives and interests, and in my opinion are a cut above those who see an unusual problem yet fail to stop and say, "Who are you? And what do you do?"

On a personal level I learned that the discovery of new information at the clinical level requires a passion for clarity and better understanding and a firm belief that with effort and diligence something not previously described or something poorly outlined can fall into sharp focus. This is a process that requires time – increasingly a fading resource – and a skill that can be taught. It requires what Pasteur, the father of the germ theory, called a "prepared mind."

The drive to discover is so strong that I suspect that we humans are genetically programmed to seek new things. Otherwise, how does one explain the actions of explorers entering oceans without maps, the uncharted pathways in Africa or even outer space travel, despite recognized hazards. Somehow the innate rewards and accolades of our peers compensate for the perceived risks. So it is with science and medicine: the dash to be the first is an enormously exciting challenge, a career-sustaining effort that tends to override frustration, rebuke, failure, hazard, political isolation, and even death as in the case of Ricketts. There may be society's failure to recognize the value of the discovery in Gram's case or worse still, the banishment from the Church as in Galileo's case when he stubbornly held to his unpopular statement that the sun and not the earth was the center of the universe.

In my own case the trip was as important as the prize: discovering or attempting to discover is energizing. Thereafter, convincing colleagues of its value is the next step. One can't wait to go to work in the morning. We know that subsequently – after publication of the outbreak – our colleagues will be able to identify similar patients and begin the best therapy. Eventually, there may be a prevention strategy developed. So it is playing the championship round that is more important than taking home the ribbon. Once we tuck the award on the shelf, we eagerly seek the next contest without much delay. People with passion and curiosity want to be a part of the Round Table. The quest for discovery is vital. It is life itself. We all seek the Holy Grail.

In the year 2000, I received a call from our third patient –then at age 57 – whose seizures 17 years later still haunted him. We had not talked since I had left Virginia in 1985. When he had learned that I had returned to Virginia from Iowa, he called me. He was preparing to visit a neurologist at Johns Hopkins hoping for relief, and wanted another copy of our manuscript that was published in 1986.

I spoke to this man again in the summer of 2001 when he told me that no drugs had controlled his seizures. Although he rarely losses consciousness, he sometimes has uncontrolled movement of his left arm and a sense of "fading in and out," a sensation of losing contact with reality. His speech stops abruptly, and he "can't get things together" when this happens. Stress seems to bring it on, and he never has a full week without such a seizure. At 6:00 or 7:00 pm in the evening, fatigue arrives like an evening fog and he is enveloped with a feeling of detachment.

He had worked for several years, returning to the same machine shop where he was previously employed before his illness, but he says "he fooled himself." He could never really do the work with any economy of time. Thus, in 1996 he went on disability. He has annual EEGs and MRIs and has tried every medication available, but nothing works. The headaches persist, and a sense of dread rolls over him. In early 2005, my patient gave me permission to use his name in this book. Now over 20 years after we became doctor and patient, I can say that Harry Cruden is alive and lucid. But surely he is not "cured." He is not "well."

There are limits to what a physician can do even with the best of intentions supported by "state of the art" science. I remind my students that 100 years from now, most of what we call modern science and clinical excellence will be considered barbaric. Even our most highly equipped ICUs will be viewed as archaic. Furthermore, uncovering the mystery of a cluster of community acquired infections surely would be little consolation to the individual victims and their families. It is a humbling reality for medicine, haunting, yet a key part of the definition and boundaries of a doctor.

DEATHS IN THE ICU
(Serratia marcescens)

Critical Care Units are very modern concepts, evolving only in the last 40 years. In fact, intravenous access for fluids and medications is only 100 years old. In the wake of the second great cholera epidemic in the early 1830s, for the first time physicians used silver canulas attached to syringes to replace fluid directly into veins in the arms. It would not be until late in the 19th century, however, with the availability of steel needles and a clearer understanding of electrolyte chemistry that the therapeutic use of IV fluids would become available. Subsequently, the advent of penicillin in the 1940s prompted the development of a still newer technology, the use of plastic catheters for continuous vascular access.

In 1978, the hospital at the University of Virginia had 700 beds, with a 16 bed surgical ICU that had an average daily census of 14 patients. Between 100 and 150 patients were admitted to the ICU each month. Trained infection control nurses surveyed all patients in the surgical ICU twice a week, seeking to identify patients with hospital-acquired infections. Every ICU patient had at least one IV line in place, almost all had an additional plastic catheter in the radial artery at the wrist to monitor blood pressure continuously, and still a few had a long vascular catheter that traversed the large subclavian veins in the upper chest cavity to enter the superior vena cava and pass through the right side of the heart into the pulmonary vasculature. One could then measure the degree of fluid buildup in the lungs and the heart's ability to move blood through the body.

During my sixth year on the faculty at the University of Virginia in Charlottesville we began to observe life-threatening bloodstream infections in ICU patients caused by a Gram negative rod with which we had no experience. In January and February of 1978, we had recorded four cases of hospital-acquired bloodstream infection due to the unusual pathogen, *Serratia marcescens*. None of the patients had had prior infection with Serratia in a wound, in the lung or urine before the organism gained entrance to the bloodstream. By definition the infections were primary, not secondary. In looking back at the surveillance data, I noted that there had been two cases in 1977 – both in November – but none in December.

Patients with bloodstream infection tend to have a fever to 102° or 103°, and an elevated respiratory rate to over 20 breaths per minute, what we call tachypnea, the consequence of bacterial stimulation of the breathing center in the brain. The heart races to a pace of over 100 beats per minute, a condition called tachycardia. If the blood pressure drops below

90 systolic – the top number – this is called hypotension, and persistently low pressure is by definition "shock." These four measures of "vital signs" are elementary and essential in monitoring such patients, several of whom developed shock and whose mortality in the next few weeks would exceed 50%.

What was surprising was that not all of patients seemed at high risk for acquiring life threatening bloodstream infection as might be expected had they all been on artificial respirators after major surgery. In fact, only some required respiratory support, yet like all patients in a surgical ICU, all had had Foley catheters to drain the bladder, intravenous catheters for the infusion of fluids, and small arterial catheters to monitor blood pressure constantly. The arterial catheters were connected to a cable that led directly to a bedside monitor with the blood pressure readings visualized in large red numbers.

I had been recruited to UVA in late 1972 just after my tour of duty in the Navy. Jack Gwaltney, the Chief of the Division of Epidemiology and Virology in the Department of Internal Medicine, wanted a new faculty member to develop research on rhinovirus, influenza, mycoplasma, and other respiratory pathogens. Jack was true Virginian, meaning that more than a few generations of ancestors had grown up in Roanoke, a two hour drive to the South and West. He was a gentleman, an equestrian who rode on fox hunts in formal regalia, and a serious duck hunter. Importantly, he had two special passions – training Labrador retrievers and studying the cause of the common cold - the rhinovirus – and he developed unusual expertise in both.

In recruiting me to the division, Jack had announced that I needed to be UVA's first Hospital Epidemiologist, a newly formed position that would demand 50% of my time. I asked him to describe the elements of the job, but he responded with some uncertainty, noting that there had been an outbreak of wound infections in patients after orthopedic surgery in the previous year. Additionally, even more critically ill patients seemed to comprise the hospital's population, and this paralleled an expansion of ICU beds. Such patients were correctly considered to be vulnerable to infections. The hospital had concluded that a dedicated infectious diseases expert should be available to respond to these problems. Importantly, and this was a punch line, the hospital would pay 50% of my salary, freeing up that amount for the department to use for other initiatives. Nevertheless, Jack had said that perhaps after 4-5 years I could focus completely on respiratory infections and hand over the hospital epidemiology reins to someone else.

In the early 1970's there were few formal programs for Infection Control in U.S. hospitals and only four hospital epidemiologists: John McGowan at Emory, Dennis Maki at the University of Wisconsin, Bill Schaffner at Vanderbuilt and I. Infections acquired in hospitals were not well quantified, and the Centers for Disease Control in Atlanta had given their measurement and control a high priority. CDC also coined the term "nosocomial" for infections acquired in hospitals. The root *noso* from the Greek means disease, and *nosokome* is a Greek word for hospital. Nosocomial infections are ones that are neither present nor incubating in patients on admission and usually become manifest only after 72 hours in the hospital.

Twenty-five years before this first assignment in my career, I had been ill with a hospital-acquired infection. No one had referred to my staph infection as "iatrogenic" or "hospital-acquired" or "nosocomial." Neither the vocabulary to describe it nor the responsibility to control such infections was developed. The concept of a hospital epidemiologist was not on the radar in the late 1940s. I came at the new task with a personal interest, and I recognized – with a sense of adventure – that it was for the most part uncharted territory.

With the help of some nurses at UVA, I had been able to design and validate an efficient surveillance system for identifying nosocomial infections at the hospital. We could perform an accurate survey of the entire hospital's population by focused reviews of the charts of high risk patients. And all of this information could be gleaned in 25 hours of surveillance time each week by the nurse epidemiologist. We could subsequently tally all nosocomial infections, divide by the number of patient admissions and calculate a rate, usually 5 to 10 per 100 (5% to 10%). Thereafter, it was possible to examine separately the infections of the urine, those at the incision site after surgery, pneumonias, and those involving the bloodstream.

A review of the specific bacterial causes might identify a cluster of infections such as an unusual grouping of *S. aureus* incisional wound infections after a specific procedure, for example a laminectory for repair of slipped discs. If all infected people had the same surgeon, and the patients of other surgeons were free of *S. aureus* infections, we had identified a key problem or risk factor. If the strains of patients had the same genetic "fingerprint" that matched the strain in the nose of the surgeon, we had microbiological confirmation and could intercede.

An outbreak, a term implying a small or large epidemic, is a high-stakes situation. The departmental faculty members on whose service the outbreak occurs are very uncomfortable and collectively hope that the cause turns out to be an environmental contamination and certainly not a clinician-related problem. The hospital administrators become anxious and have three substantial fears: adverse publicity, a warning from the Joint Commission that performs hospital accreditation, and the dreaded lawsuits. Nurses are fearful that whatever the cause, they will be the initial focus of attention. The spotlight is also on the hospital epidemiologist to solve the mystery, to end the problem quickly and quietly without incident, and hopefully to implicate a cause apart from the staff, perhaps the patients themselves. If the solution is more evasive than anticipated, it must be that the hospital epidemiologist – or more frequently at that time the nurse epidemiologist – is not up to the job. While each of us is concerned that a poor outcome might reflect on ourselves or our professional group, all of us worry about the patients and have as our basic charge to do the best in their interests.

There are three key triggers to identifying a problem early: a proactive surveillance system, a curious microbiology technologist or astute clinicians. Any one or a combination of early warning alerts could be valuable. If the rates of infection identified by reliable surveillance rise to a level rarely seen before, then an outbreak is identified. If a curious lab technician sees a cluster of infections associated with an unusual bacterium at the same anatomical sites, he or she might alert the hospital epidemiology team. A physician also might notice a larger than expected number of infections with the same organism occurring in a tightly defined time and location.

The newly appointed director of the Surgical ICU was John Hoyt, an astute clinician trained in Anesthesiology and Critical Care Medicine. He asked me a question about managing one of the patients with Serratia bloodstream infections. He knew that the most severely ill patients, often with the highest risk of hospital acquired infections, were found in his critical care unit. Furthermore, because so many devices are employed in monitoring their vital signs and physiological responses to drugs, those in the surgical ICU are especially at risk.

John was short, stocky, thoughtful and generally soft spoken. In trying to characterize his personality into the dichotomy of either introverted internist or extroverted surgeon, he was clearly more the internist. In beginning to discuss the new infections, neither John nor I had previous experiences with Serratia infections.

Even now, thirty years after my original encounter with these bacteria, *Serratia marcescens* is still not a household phrase. This is remarkable in light of the legendary history of the organisms, named in 1823 by an Italian pharmacist, Bartolomeo Bizio. Members of this species have a predilection to grow on foods rich in starch like bread, utilizing the starch for energy after appropriate metabolism. Many isolates produce a red pigment, and for over 2500 years there had been stories of "blood" magically appearing on bread.

In the summer of 1819, in Northern Italy there were alarming reports of red spotted polenta, a popular corn meal product. A commission was appointed to investigate this strange occurrence, with Dr. Bizio as one of its members. He offered the unusual hypothesis that the seeds of a microorganism caused the discoloration, and not divine intervention. Specifically, he said that a "fungus" had caused contamination of polenta. To test his idea, he placed normal appearing polenta at constant temperature in a damp environment. After a few days he observed some initial spotting of the corn meal with red color and later a blanket of red pigment entirely covering the polenta. He concluded that his hypothesis was correct. It was later realized that a pigmented bacterium rather than a true fungus caused the contamination.

Bizio named the genus after Serrafino Serrati, a notable Italian physicist who had developed a steam engine. Apparently Bizio thought that too much honor had been bestowed on Americans who had overlooked the important work of Serrati. The word *marcescens* is derived from the Latin meaning to dissipate, and it was meant to convey the progressive fading of the bacterial pigment production after repeated subcultures. In fact, not all species produce a red pigment, including those invading our ICU.

At UVA one of the young fellows in training with me was Leigh Donowitz, a petite, slightly built and incisive Pediatrician and Infectious Diseases expert. She had had a special interest in nosocomial infections acquired in critical care units, having migrated from Philadelphia for both residency and subspecialty training in Infectious Diseases. I asked Leigh to review the charts of all infected patients, make a list of the dates of ICU admissions and infections, define the average interval, look at all physicians involved, prior procedures, prior medications and all bed locations within the ICU. If new cases occurred, she was to keep adding the information to the hand written charts. She would also look at the intervals between medications and infection, operations and infection, contact with specific physicians or nurses and infection, and all ICU procedures and subsequent infection. With meticulous detail Leigh developed this useful database in the pre-computer era of the mid-1970's. She would show that the rates of Serratia bloodstream infections were statistically higher than expected by chance alone. We had an epidemic!

John Hoyt had both curiosity and tenaciousness. He also had an additional motivation to solve this problem as an anesthesiologist in charge of a surgical ICU. Feeling the day to day tension between the two disciplines of Anesthesiology and Surgery, he knew that if anything adversely happened to the patients while the surgeons were in the operating room, accusing fingers would point to him. In some hospitals surgeons managed their own ICU patients, but there was an increasing body of knowledge that well trained critical care experts who did this full time had better patient outcomes than those who did this only part time.

We all became alarmed in March when six more cases of Serratia bloodstream infections occurred, and we could find no obvious connection to a cause. In April, seven more cases were identified, and we were close to panic mode, not just because the numbers were escalating, but because patients were dying. Ten of seventeen patients with primary *Serratia marcescens* bloodstream infections (59%) would eventually die in 1978.

As the attending physician of record, the individual surgeons whose patients were in the ICU had the painful task of telling the families about the infection that led to their demise. John Hoyt was very helpful and sympathetic. Some of the 17 men and women had serious underlying diseases and obviously were aware that there are risks to surgery and hospitalization in critical care units. None, I am sure, ever imagined that they would die as a victim among several in a new epidemic affecting hospitals in many cities of the United States.

Only a few decades earlier, Serratia were assumed to be harmless bacteria, and microbiologists took advantage of their pigmented nature to study infection. For example in the 1950s, our government sponsored the release of Serratia into the air off the coast of California and in the subways of New York City to examine the feasibility of potential biological terrorist attacks. Some patients claimed that infections followed those experiments, but the discussion is controversial. Other investigators placed the red organisms on the outside surface of Foley bladder catheters to track their travel time to the bladder. Clearly in the mid-1970s, there was not a full appreciation that *any* organism can cause disease in the right host, especially if it enters the bloodstream of patients. Our data would attest to the ability of Serratia to kill.

We had cultured the Foley bladder catheters, the respiratory tubing of those patients who were intubated, and the wounds of those who had had recent surgery. Swabs for culture were obtained from patients' bedside tables, a surrogate for environmental contamination. Serratia was nowhere to be found. We cultured the blood pressure cuffs, and our swab samples failed to yield the organism.

We had cultured the fluid traversing the IV lines of 51 patients, recalling an outbreak of bloodstream infections in the late 1960's and early 1970's that had been traced to a manufacturing site of intravenous fluid where the tops of the IV bottles were contaminated by nearby water. No Serratia.

We then sought to culture the hands of ICU personnel in case some transmission was occurring from patient to patient. To avoid any immediate embarrassment among ICU personnel and to enlist their full cooperation, we chose to pool the cultures and run them anonymously. Two or three hand cultures would comprise each pool, and the combined collections were cultured. We had eight pools, and our first break occurred when one of the pools was shown to contain Serratia.

Why would Serratia be on the hands of health care workers when we couldn't find it anywhere in the environment? If hand transmission was important, how did the organism get from the hands of nurses, physicians or technicians to the bloodstream of patients? Where was this organism usually residing, its home location, what we call its "reservoir"? Should we close down the operation of the ICU?

Just then a new lead surfaced from the epidemiological data in the spring of 1978. Leigh Donowitz had finished examining the timing intervals between infections and prior "exposures" to specific medical milestones in their care. The data showed that the intervals between exposure to IV lines, medications, specific physicians or nurses or most procedures and the subsequent Serratia infections were erratic and no theme emerged. However, all patients with Serratia had had arterial catheters placed in a narrow time interval of 2-3 days before the bloodstream infections.

At the next morning meeting of our team with the ICU staff, we reviewed all new policies, procedures and equipment introduced to the ICU in the last year. We told everyone that there appeared to be a relationship between the placement of arterial catheters and subsequent Serratia bloodstream infections. One of the ICU nurses said that the only new procedure was the introduction of a pressure monitoring device attached to the arterial catheter apparatus with improved safety features, in fact introduced to *reduce* ICU related bloodstream infections.

In some previous outbreaks investigated by the Centers for Disease Control in 1976 and 1977, organisms were found on transducer heads, the pressure sensing devices at one end of the cable that connects directly to the intraarterial catheters. The cable at the other end of the transducer led to the blood pressure recording apparatus. A new disposable piece of equipment that wrapped around the transducer head had an impervious plastic barrier – polycarbonate – between the transducer head and the catheter entering a patient's artery. This was designed specifically to prevent any organisms, possibly residing on the transducer head, from entering the catheter leading to the arterial blood of patients. Its shape was like a cathedral's dome, and the disposable "dome" became a necessary purchase for all ICUs.

A microbiology report a few days later jolted us: one of our newly admitted ICU patients was found to have *Serratia marcescens* meningitis but with negative blood cultures: the spinal fluid was infected but curiously not the blood. We all three rushed to examine the patient who had had a recent neurosurgical procedure. At the bedside we noted the pressure monitoring apparatus with the new disposable dome. But this time it was not engaged with the patient's artery, as it would have been for the rest of our patients in order to monitor their blood pressure. In this case the neurosurgeons were seeking to avoid any undue pressure buildup in the brain post-operatively and decided to monitor the situation continuously. The catheter connecting to the transducer and dome was placed below the skull and into the spinal fluid that is in close contact to the tissue surrounding the brain, the meninges. In our minds, it was now clear that the pressure sensing device was directly linked to infections.

Light bulbs went off, and we cultured the transducer heads of all eight patients being monitored in the ICU. The following day we gathered in the microbiology laboratory where we heard the news that all eight were positive for Serratia. We now had microbiological confirmation of our epidemiological information linking transducers to infection. The immediate reservoir for Serratia in our epidemic was the transducer heads of pressure monitoring equipment. The bloodstream infections had resulted from organisms that gained direct access via the arteries, and the single case of meningitis resulted from Serratia's migration directly to the spinal fluid and meninges via the pressure monitoring catheter. Immediately we introduced a policy to place disinfectant on the tops of the transducer heads routinely to end the outbreak.

Could the organisms residing on the transducer head be penetrating through the plastic barrier? If so, that would have major impact. A technology company manufacturing the dome may have created a flawed product that was putting thousands of ICU patients around the country at risk of death. We physicians had uncritically accepted the thesis without question that technology had succeeded in reducing transducer-related infection. However, we now had to ask if organisms residing on the transducer heads may be passing through unsuspected pores of the polycarbonate plastic "barrier."

Leigh, John and I decided to do an experiment in the lab with Fred Marsik, our clinical microbiologist: we seeded high concentrations of organisms on the transducer heads, squeezed the dome in place by screwing it on the metal threads as tightly as possible and waited to see if organisms that we had placed on to the head could break through the membrane to the "patient side." All of our microbiological attempts failed to support the hypothesis.

The epidemic was quickly aborted with routine disinfection of the transducer heads, confirming the critical hypothesis that the reservoir for infections was the transducer heads. A nagging question remained, however: how were the organisms getting from the transducer heads to the patients' bloodstream? We then remembered that we had a single pool of hand cultures positive for the organism, suggesting contamination of hands from contact with the transducer apparatus. John Hoyt arranged for us to view the details of how the transducer was assembled and used in patients. He also brought a camera.

An ICU technician illustrated the procedure, first showing us the transducer cable with a plug into the monitor on one end and the transducer head on the other end. In order to achieve optimal transmission of blood pressure from the patient's artery via the catheter and then via the transducer to the digital read out monitor, it was necessary to apply a small volume of sterile saline to the transducer head before screwing on the disposable dome. A meniscus of fluid was needed. We watched as the sterile disposable dome was then opened from its wrap and screwed around the threads of the transducer head. The apparatus now assembled, was mounted on a stainless steel pole by the bedside.

John was the first to see the clue, and we have documented it on film. The space between the index and second finger of the person assembling the device was glistening with a thin layer of fluid. To understand how the saline fell onto the fingers, imagine the following analogy: you are holding a large bottle cap with a diameter the size of a fifty cent piece. The top is against your fingers upside down with the open end pointing upwards. You then place a volume of fluid to fill up the cap, and there is a small mound of fluid above the edges. You then try to fit a dome shaped piece of plastic over the sphere of fluid and within the cap. But the dome is smaller than the volume of the bubble of fluid, and in the squeezing process, some fluid falls around the cap and onto the fingers.

During the process of assembling the device in the ICU, the fluid couple on the transducer head was too voluminous for the space between the transducer and the dome's capacity, and the excess had fallen onto the fingers of the person assembling the device. The saline was indeed found to be originally sterile but when it bathed the contaminated transducer head, it had become a vehicle for Serratia.

The next step in the procedure was to touch the 3-way stopcock attached to the arterial catheter, turning the plastic lever to change the flow of blood away from the outside port and now direct it to the transducer. One arm of the stopcock leads directly to the column of blood in the artery. This is very helpful for drawing blood for chemical tests, especially to measure arterial oxygenation or acid-base balance. However, direct finger contamination of the port leading to the artery allowed Serratia access to the bloodstream. The organism was cultured from the transducer heads – that was its immediate reservoir. In fact, *Serratia marcescens* did not penetrate the defense wall of the dome's impermeable membrane directly to the arterial blood. It made an end run around the dome on the fingers of those who assembled the pressure monitoring equipment at the bedside.

The implications of these observations were stunning: modern technology was designed to solve a problem, transducer heads contaminated with bacteria, by developing a truly impenetrable membrane barrier between the apparatus and the patient. But in the day to day assembly of the product, fingers become contaminated, and routine human behavior in the ICU had led to an unexpected epidemic. What many had assumed to be a surefire cure, in fact a triumph of technology, actually caused the problem.

We began disinfecting the transducer heads with a liquid disinfectant called glutaraldehyde routinely eliminating the organism on its reservoir, and the outbreak never recurred. Soon thereafter we prepared an abstract for our national meeting and began a manuscript for publication in the Journal of the American Medical Association, *JAMA*. We also strongly suggested to the manufacturers of the dome that they add a new insert with the directions for routine disinfection of the transducer head whenever using the apparatus.

In an attempt to prevent similar outbreaks, we also wrote to the Food and Drug Administration outlining our concerns about the need for a glutaraldehyde disinfectant even when employing the new plastic dome. A representative of the FDA came to UVA eight months later and did nothing. Moreover, the manufacturer of the dome refused to print any suggestion that disinfection was needed in using their product and hinted at a lawsuit if we published the article. Our paper in *JAMA* was published on October 19, 1979, and we heard nothing further from the manufacturer.

Here was another lesson. Unless the agencies in charge of monitoring systematic shortcomings in medical products respond not only to flaws in the product but also to flaws in how it is to be used, technology will frequently be trumped by human behavior in hospitals and clinics. With lives at risk the response needs to be timely. Obviously the manufacturer took no responsibility for the in-use performance of the product, only for its specification if used flawlessly by health professionals, with handwashing between every phase in setting up the transducer.

Perhaps more than any other event in my early career, this epidemic cemented my interest in preventing and controlling nosocomial infections. The detective work involved in controlling an outbreak was compelling, and the excitement of managing the problem in as brief a time period as possible was an adventure. This was front line action in Infectious Diseases. There was tension to solve the problem quickly but also the rewards of success at the end if the work went well. Unlike the clinical venue where intervention can have a favorable impact on a single individual, effective epidemiological work can improve the lives of large populations. Surely future infections and deaths were prevented, and I was committed to the field of Hospital Epidemiology.

Serratia would return in the form of déjà vu in the Spring of 1986 while I was on sabbatical in London. A good friend and colleague from Brussels, Professor Jean-Paul Butzler, the Chair of Microbiology at St. Pierre's University Hospital, asked me to give a few talks on the epidemiology of infections acquired in hospitals. Even in the mid-1980s, he himself was already recognized as an accomplished scientist, having discovered an organism causing gastrointestinal infections, Campylobacter. In Europe and the United States, this is the leading cause of infectious diarrhea.

Like many Europeans living on the continent, Jean-Paul had outspoken disdain for the rainy weather of London and wanted to show me the special climate of Belgium. Previously I had concluded that the weather in Belgium and London was quite similar. But he insisted, "Bring your wife and children to our home in Knokke on the coast. If it rains one day, we can excuse ourselves to Brussels for a day of meetings and return to the families that night."

All four Wenzels crossed the English Channel from Dover to Knokke via Oostende on the Hydrofoil, arriving to magnificently sunny and warm weather in May. The sandy beaches were a great contrast to the wet streets of London, and Jean-Paul and his wife Rolande were generous hosts, introducing us to fine wines, savory French cooking, and some specialties such as beef tartare, smoked eel, and champagne moules, the most magnificent mussels in Europe.

By day three I was anxious to keep my end of the bargain, but the weather remained perfect, and Jean-Paul insisted that we remain in Knokke. That evening, however, he received an anxious call from a former trainee, who was in charge of the microbiology laboratory at a modern 700 bed hospital in Belgium: "Would you and Dr. Wenzel review our investigation of over 70 cases of *Serratia marcescens* bloodstream infections in the

75

ICU." I was pleased to do some work for my friend, and we would arrive for a formal briefing in the morning.

In Medieval Europe there had been numerous reports of "bleeding" of the Catholic Church's sacramental hosts, the pressed wafers fed to parishioners and believed to represent the body of Christ. With gross prejudice, Catholics at the time ascribed the bleeding to a defiling of the hosts by Jews. There were even rumors that Jews had stabbed the hosts, and as a result Jews were executed by the thousands. In a recent scientific report, Dr. Leo Greenberg described the story of the chief rabbi of Brussels who in the 14th century was alleged to have purchased host wafers from the chapel of St. Catherine, stabbed them and caused bleeding. As a consequence of unbridled hatred, the Jews of Brussels and surrounding towns were summarily executed. From these stories it was obvious that *Serratia marcescens* was no stranger to Belgium and had previously contributed to the premature deaths of innocent people.

In the hospital's conference room the director of the ICU formally reviewed the data for all those attending: the ICU physicians, surgeons and nurses, the microbiologist, Jean-Paul and me. More than 70 cases occurred in the last 12 months with over a 25% mortality. All environmental surfaces in the ICU and all water sources were negative for Serratia. The microbiology studies of patients were negative including throat, urine, wound site, sputum and rectal cultures. The IV fluids were cultured and found to be negative.

In deference to me, everyone spoke English, an easy feat for physicians in Belgium. I inquired if patients had transducers and arterial lines in place with plastic domes.

"Yes," was the answer.

I could hardly contain my excitement at the probability of pinpointing the cause of a year long epidemic within one hour.

"Did you culture any of the transducer heads?," I asked.

"No," was the response from the physician director. He cited this brusquely as though it was a ridiculous idea.

"Well, I think you will find that your source for Serratia is on the transducer heads," I volunteered.

"Very unlikely!" his quick retort. This time I realized that he was upset with my suggestion.

I tried a new approach, "When you assemble the arterial monitoring device, do you need to use a fluid couple on the transducer head?"

"Yes, we do," said one of the nurses softly.

"Sometimes fluid from the transducer falls onto the fingers, and after mounting the apparatus, does the nurse or physician touch the 3-way stopcock directly in contact with the arterial blood?" I added.

The ICU nurse nodded affirmatively.

I volunteered to help culture the transducer heads, but the team demurred.

After a brief tour of the ICU, many smiles and perfunctory handshakes, Jean-Paul and I left the hospital and went to a nearby restaurant. Over lunch, I began to second-guess my approach at the hospital and regretted not making the solution appear more opaque. I could have been Socratic in my consultation, offering suggestions that would lead the local team to the answer. I realized that I showed modest cross cultural tact, and certainly no international diplomacy. After all, the equivalent of tens of thousands of dollars were spent at the hospital for thousands of microbiological cultures. Hundreds if not thousands of hours of effort had been expended, and an overconfident American had come to a brief meeting and blurted out the only piece of equipment that was not cultured, suggesting that this was the missing link. In the pre-web era of the 1980s, it was more difficult to scan the literature for articles on Serratia bloodstream infections in ICUs. An additional problem in the 1970s and 80s was the bias of many physicians to read the literature only in their own national journals and language. Internet technology would arrive in the 1990s and flourish by the end of the 20th century, and within minutes one can now find the pertinent references in the published literature of many languages. But that was not happening in most hospitals in the world in the mid 1980s.

A week after my visit to the hospital, I was giving a paper in Berlin, and in the audience was the microbiologist who had asked for my help. He told me that after a new case had occurred, he had cultured all of the transducer heads in use in the ICU, and they all grew Serratia. He was effusive in his thanks and relieved that his outbreak would end. Disinfection of the transducer heads was begun, and so far there had been no new cases. Even a year later there were no new cases. We have remained as friends in contact for the subsequent 19 years, meeting at annual Infectious Diseases conferences. In the spring of 2001, he invited me to Brussels to be a speaker at a special symposium honoring the career of Professor Jean-Paul Butzler.

Nosocomial or hospital acquired infections occur in 5-10% of patients entering U.S. health care institutions. The majority currently cannot be prevented, but perhaps 20% are fully unnecessary. Most occur sporadically, and a large proportion occurs in critical care units. In some situations, as I learned early at UVA, infections occur as part of an epidemic.

Epidemics by definition represent preventable disease, and they occur more commonly than many recognize. Perhaps 5% or more of nosocomial bloodstream infections in U.S. hospitals occur as part of an epidemic or small cluster. They are challenging and threatening both to personalities and to institutions. In the face of initial ignorance and uncertainty, irrational finger pointing is common, and when it comes to explaining causes, there are enormous biases from irrational, preconceived ideas. Such prejudice results sometimes in deadly consequences. With time, data and insight, the truth emerges.

The experience with Serratia illustrated a number of lessons about technology and bias. In contrast to random error, defined as a chance phenomenon, bias is in fact a systematic error in thinking. When applied negatively to a population of people, we call it prejudice. When some frightening event occurs that defies explanation, many people rush to blame others, the outsiders, the ones who are different. The history of the red pigmented bacteria is riddled with such examples, and our experience with societies would say that such bias is not rare.

We may also be biased about readily accepting credible ideas from outsiders. We might all admit that feeling, and I experienced it initially in Belgium after my quick explanation of an epidemic. It is uncomfortable for us to admit that we have worked hard for an elusive answer or a solution to a problem, when a "different" person has one at hand. I also learned that for a person to be a helpful consultant, one needs to lead the requesting group to the answer that they discover as a result of your questions. All good teachers know this.

As individuals and as societies we even have biases about the wonders of technology: it will solve our needs and be forgiving of human behavior. But all technology needs to be questioned and closely observed after its implementation. We cannot accept all of its exciting promises blindly. The irony of the transducer problem is that a "solution" led to an unexpected widespread problem.

In the future there will be new technology, beautiful profiles of its performance, and new biases. New problems will emerge with serious consequences. Hopefully, we will do our best to be rational, to learn from past experiences, to keep our minds open for surprises, and to greet foreign solutions without prejudice.

As one examines the characteristics of a doctor with the lens of experience, more refined feature become apparent, and an earlier vague outline is beginning to take on a sharp focus. The quality to anticipate surprises comes into view with a parallel need to avoid bias, to meet the unknown with equanimity. This is critical for the benefit of individual patients and for populations of patients at risk for adverse outcomes due to specific vulnerabilities common to all of them. If there is a lesson for those of us committed to training future physicians, it is this: the greater requirement in medicine is to challenge our assumptions, not our observations.

IN THE SHADOW OF
SEMMELWEIS
(Enterococcus, Proteus mirabilis,
Streptococcus pyogenes)

In Iowa I received a call from a colleague working in a private hospital in Pennsylvania asking urgently for help with an outbreak of vancomycin – resistant enterococci, Gram positive spherical organisms. Infections were occurring in the bloodstream of patients with underlying leukemia, and the mortality was high.

The critical reason that patients with leukemia suffer from life-threatening infections is that the disease is associated with extremely low numbers of neutrophils, the major white blood cell type that engulfs bacteria and kills them. Furthermore, chemotherapy for leukemia often causes an inflammation of the gastrointestinal (GI) tract, the reservoir or home of enterococci. As a result of the loss of integrity of the mucous cells lining the intestine, bacteria manage to find their way to the bloodstream. The effect is that the threshold number of bacteria needed to cause infection is much lower than that usually required when white cell counts are normal and there is an intact GI tract.

I asked Mike Edmond, at the time my fellow in training, if he would be willing to travel. Mike, a slim young physician from West Virginia, has a facile mind and perennially, wide-open eyes as though on special alert. He listens so attentively that one senses the gears of his mind quickly grinding out solutions. He is a tireless workaholic and a true night owl, his favorite time for working. In a few words, he is intense, talented, and tenacious. I explained to the caller that Mike was soon to be a graduating subspecialist with extraordinary skills in microbiology, infection control, epidemiology and computer support. In fact, Mike is a rare physician who is comfortable with the unknown and enjoys challenges. He is unusually perceptive yet loves numbers and spreadsheets. He uses both right and left sides of his brain with ease.

Consulting with the community hospital, Mike reviewed the outbreak, visited the wards where patients were housed and began to compare the infected patients to those fortunate enough not to have been infected. This approach is called a case control study, one in which the object is to identify "exposures," the specific experiences that were more common among the cases before the infection occurred, than among the controls.

If a particular experience is statistically linked to cases, it is called a "risk factor" for the bloodstream infections.

When initially identified, risk factors are only associations. It takes other information such as bacteriological studies to confirm cause and effect relationships. Mike identified two important associations that linked our cases to bloodstream infection. All cases had been colonized with the same antibiotic-resistant organisms in the gastrointestinal tract for about one week prior to the subsequent bloodstream infection; and all cases had received antibiotics that target anaerobic bacteria, the many species that survive in extremely low oxygen environments found in the colon. It is generally thought that anaerobic bacteria for the most part are protective, keeping their aerobic neighbors in check – *E.coli*, Proteus, Enterococci and others.

In contrast to the experience in cases, none of the controls had been colonized with vancomycin-resistant enterococci, and only a few had received anti-anaerobe antibiotics. The statistics showed that the probability that the differences observed could have happened by chance alone was well under 1%. The threshold that we accept – by convention – as statistically significant is 5% or less. In scientific journals this is recorded as the probability (p) is less than 5% or ($p<0.05$).

One of the great sacred cows of science is the acceptance of the notion that the "small probability" that a difference exists should rest at the 5% level. I examined the origin of this idea a number of years ago, anticipating both a cumbersome mathematical formula and an arcane discussion. I was surprised to find that Sir Ronald Aylmer Fisher (1890-1962), a brilliant British statistician well known in the early half of the 20th century, had been asked a question by English farmers: How will we know, if we add fertilizer to our crops in the field, if an increase in yield is causally related? His simple response: If you've never seen a crop that size in 19 out of the last 20 years, it represents a relationship that is not due to chance alone. That is the origin of the very arbitrary, 5% probability level of "significant" difference that all sciences accept in comparing tests of new vs. old treatments. Fisher, with great wisdom, made it up! - and I tell my students – with tongue in cheek - that it all goes back to horse manure.

Two important risk factors in the study of antibiotic-resistant, Enterococcal bloodstream infections among leukemia patients were now identified. However, there were a number of biological questions posed by the new information. Were the risk factors causally related to the bloodstream infections? If we asked the clinicians to stop the use of all antibiotics that target anaerobes, would that reduce the infection rate? How were the antibiotics causing the infection or facilitating it in any way? Could we do anything to stop or suppress the growth of enterococci living in the lumen in the GI tract, which appeared to antedate the life-threatening bloodstream infections? Were all the organisms the same, and if so, how were they moving from patient to patient?

The infection control implications were straightforward: First, allow no more prescribing of antianaerobic drugs for the patients. Our hypothesis was that antibiotics directed at the anaerobes were creating a competitive advantage for the aerobic vancomycin-resistant enterococci. By altering the balance of power in the intestinal tract, the enterococci were growing to such high numbers that they were overflowing into the bloodstream highways by a process that still was uncertain.

Secondly, handwashing had to be assiduous to prevent transmission from patient to patient, and preferably with medicated soaps, shown earlier to be more active against the enterococci than non-medicated preparations. Although exhorting healthcare workers in critical care units to wash hands appropriately has been unsuccessful anywhere, compliance with handwashing on leukemia and cancer wards has been less problematic. I think that the nature of the underlying diseases and the closer relationships between nurses and patients on oncology units are especially important influences.

Eventually, genetic fingerprinting of the antibiotic-resistant organisms showed clonality: all patients were infected with a single strain of Enterococcus. Surveillance, statistics, and genetic typing showed that we had an epidemic with a single strain of enterococcus spreading from patient to patient. Fortunately, handwashing compliance plus new antibiotic prescribing restrictions were linked to control of the outbreak, verified by ongoing surveillance. From a species interactions perspective, an organism of low virulence became a deadly pathogen in patients by infecting hosts with limited ability to ward off infections. In simplistic fashion, infection risks increase directly with the number of organisms to which a person is exposed, the increasing virulence or innate aggressiveness of a pathogen, and inversely with host resistance:

$$\text{Infection} = \frac{\text{Dose x Virulence}}{\text{Host Resistance}}$$

Furthermore, the Enterococcus' ability to acquire easily transmissible genes for vancomycin resistance gave it a selective advantage in modern U.S. hospitals, where vancomycin is used in great quantities. Although no drugs were then available to suppress enterococcal carriage in the intestinal tract at the time, I would later have the opportunity to study one with suppressive ability after moving to Richmond.

Vancomycin was considered the last available drug with known activity against the enterococcus, but we now face vancomycin-resistant enterococci—VRE. Mike Wong, a new Harvard-trained recruit to our faculty led the effort, and in a multicenter study we found that a new oral antibiotic called Ramoplanin would markedly suppress the growth of VRE during the time that patients took it orally, and for one week afterwards. Beyond that period, the colonization rates were back to baseline. Our study was published in *Clinical Infectious Diseases* in November 2001. Imagine the earlier outbreak involving vulnerable patients at a hospital in Pennsylvania if we had had a drug to suppress the VRE carriage in these patients with leukemia for 2-4 weeks – the period when their while cell counts were depressed by the chemotherapy. It is possible – but not yet proven – that we could have prevented some of the bloodstream infections and deaths.

In the meantime, the solution to preventing further enterococcal bloodstream infections at the hospital in Pennsylvania was uncompromising handwashing. However, the key to understanding the mechanism by which the enterococcus became resistant to all antibiotics is to realize that microorganisms are promiscuous creatures, their unbridled sexual appetites leading to a rapid spread of antibiotic resistance genes. They weren't always that way but seemed to develop a free love attitude only after the introduction of antibiotics.

In the preantibiotic era of the 1920's and 1930's, there were some populations of bacteria that had low frequencies of resistance to ampicillin and tetracycline even decades before these drugs were commercially produced. We know this because organisms collected and saved during this period could later be tested against the newly developed antibiotics. Two to nine percent of large collections of *Salmonella*, *Shigella*, and *E.coli* expressed resistance, and one would surmise that because soil contains molds that manufacture antibacterial substances, it should not be surprising that bacteria with soil exposure would show a low level of resistance. Note that these Gram negative rods are sometimes part of the intestinal flora of animals, and soil exposure would be common. Importantly, the genes coding for resistance were located only on the bacterial chromosome, the DNA residing in the nucleus of the cell. Most of the common organisms causing disease would be susceptible to these antibacterial products.

In Northern Africa there is a town called Wadi Halfa where an ancient tribe of Nubians, the X-Tribe, has been the focus of study by anthropologists, in part related to their having had unusually low rates of infection that caused death. Post mortem examinations have not shown the tell-tale signs of infection at the same rates as other primitive populations studied. With modern technology it is now possible to do studies on bone, and the bones of buried members of the X-Tribe from 300 AD have been shown to contain remnants of the antibiotic, tetracycline. This observation is not of the result of recent contamination but of ingestion at the time of their living. The main staple of the Nubian tribe members was a grain that was easily contaminated with fungus during storage, the same fungus that naturally manufactures tetracycline. The grain was used as an ingridient both for bread and beer, and it is likely that the ingestion of the foods and drink prepared from the grain had a favorable effect on simple infections of primitive populations over a thousand years before Sir Alexander Fleming discovered penicillin and heralded the antibiotic era.

Microorganisms deserve our respect for their ingenuity. With the advent of the clinical use of antibiotics, some of the genes for resistance to many antibiotics jumped from the nucleus to the cytoplasm, where they engaged the circular forms of DNA called plasmids. The significance of plasmids is that they are readily shared by consenting bacteria in close proximity, and important genes are transferred in the process. Imagine two organisms of the genus Enterococcus that usually inhabit the gastrointestinal tract. If one has a plasmid in the cytoplasm with a gene coding for antibiotic resistance to vancomycin and the other has no such gene, the following ritual can occur. The recipient sends out pheromones, only 7 or 8 amino acids in length, that are detected by the donor. Now excited, the donor organism will increase the baseline frequency of gene transfer by 100,000 to 1,000,000 fold! So if the background rate were gene transfer of one time for every 100 million bacteria, it could rise to 100,000 to 1 million per 100 million organisms. The donor accomplishes this task by organizing a protein-based, connecting bridge. Once the gene for resistance is transferred, the original recipient shuts down the production and release of pheromones.

When I discuss this with students, I suggest that if the donor organism were interviewed and asked about the affair, it would have this to say: "I did not have sexual relations with that organism, the enterococcus!" I add quickly that, whereas this meets the legal definition, it fails to describe everything that transpired. In the antibiotic era, with genes for resistance on plasmids, a high prevalence of antibiotic resistance can occur rapidly in large populations that are separated widely in space. The enterococci causing infections on a leukemia ward in Pennsylvania had genes that coded for resistance to the antibiotic, vancomycin.

For decades, the enterococci were the Cinderella's of Infectious Diseases, having received little respect in our field. Occasionally after bowel surgery, these organisms would be found in cultures of an abscess or wound but always in the company of other organisms thought to be the true "cause." Enterococci were considered to be accidental tourists with little virulence yet frequently in the company of sister organisms, *E.coli* or Bacteroides – the truly "important" Gram Negative rods. Many physicians argued that they need not be treated. By 2005, however, they wore the glass slipper for pathogens causing nosocomial infections. The Gram positive bacteria that form chains rather than grape-like clusters are the third most common cause of nosocomial bloodstream infections in U.S. hospitals. The mortality for the 35,000 patients who acquire this infection each year is 30%. Unfortunately, 25% of these organisms are resistant to the last chance antibiotic, vancomycin. These estimates are based on data from ongoing surveillance of infections in hospitals.

When I first accepted a position as hospital epidemiologist in 1972, I had no notion of what priorities I should address. It was a natural idea for me to think that if I knew what the most important problems were, I could channel my efforts into solving these. I needed to know what the leading organisms were that caused nosocomial infections and what antibiotics were effective for their treatment. It was a quick leap to the idea that some type of ongoing survey of hospitalized patients would give me the rates of specific infections acquired during their stay. Perhaps there was a way to identify patients at high risk for infection, and I might also describe the impact of infections – how many died or required extended hospital stays. I needed a surveillance program.

I had not realized it in the early 1970s, but this was the beginning of excellent epidemiology: defining the rates and distribution of disease in populations. Previously a few hospitals collected a series of cases with specific infections, but none had published a method for ongoing estimation of rates. Rates required a numerator and denominator: of all people at risk (denominator), how may acquired a specific infection (numerator). Epidemiology begins with long division.

There are several reasons why incidence rates of infection had not been developed as a result of a surveillance program. No one knew how to make it efficient and accurate, and it can be tedious work for the surveyor. Another reason delaying the broad acceptance of surveillance I would learn after being invited to give a formal presentation to the Department of Surgery at UVA. When I was introduced by the senior surgeon in charge of Surgical Grand Rounds, he dutifully read my credentials as hospital epidemiologist in charge of controlling nosocomial infections, adding that

I was the new "nosey communist," a derogatory reference to my viewing the quality of their care and a play on the word "nosocomial." Obviously I was intruding on their turf and might identify problems that occur within their domain, none of my business. It is common for bright, well-intentioned people to make erroneous estimates of the extent of a problem, both high and low, because of their exposure to a limited portion of the hospital or at a ward. They may also have biases that lead to a denial of the extent of a problem. In summary, surveillance and reporting were not part of the culture of academic hospitals in the early 1970s.

In order to identify some extraordinary problem – an unusually high rate of epidemic of antibiotic-resistant infections – one had to know the background rate of expected infections, the endemic rate. Surveillance would be essential for this distinction.

Antibiotic resistance, which is coded by a gene on the chromosome or on a plasmid, can be a crude fingerprint of a species. Thus, if four patients develop a staph infection after spine surgery and three of four have an organism showing resistance to the antibiotic, methicillin, we might initially discard the case caused by the susceptible organism. Thus, we don't have to look at the rates infections of all *S.aureus*, and instead we can focus on the subset of methicillin-resistant *S.aureus* (MRSA). Additionally, we learned early that rates of methicillin-resistant *Staph aureus* are a good surrogate measure of handwashing. In hospitals with poor infection control, rates tend to be higher than in those with better handwashing compliance. A focus on MRSA is good for clinicians and epidemiologists alike, because the former are concerned about treatment options in individual patients, and the latter have a marker of these especially troubling bacteria. Furthermore, with follow up we could trace the fall in rates (hopefully) after the introduction of specific measures to control the spread of infections in hospitals. The introduction of effective surveillance would lead me on the right path.

The first infection control nurse with whom I worked was Chuck Osterman, a man who migrated from ward nursing to infection control in 1970. He was one of the country's initiates in the fledging field of hospital infection control.

Imagine the task: you have 600 patients in the hospital at any one time at risk for infection and the beds turn over every seven days on average. How does one nurse epidemiologist survey the hospital each week with sufficient detail to get accurate rates. The answer was right under our noses, and I was fortunate enough to see it. The nurses on each ward kept a single card – the kardex – with each patient's name, age, date of admission, diagnoses, operations, and every invasive device such as IV line or foley bladder catheter. It made sense to me that patients with foley bladder catheters were more likely to get a nosocomial bladder infection than those without; those who have surgery are the only ones at risk for an incisional wound infection; patients in the hospital for fewer than three days by definition cannot have a nosocomial infection; those in ICUs are highest risk patients for nosocomial bloodstream infections and pneumonia. We set up various lists of high risk patients, reviewed the kardex once a week on general wards and twice a week in ICUs. We then would review the complete charts only of the high risk patients and not open the charts of other patients. We then proved that this type of screening was accurate and took relatively little time compared to looking at all charts daily.

Having developed efficient and accurate techniques to identify serious hospital acquired infections, Chuck and I began teaching this to newly designated infection control nurses. This experience led to our expanding the UVA approach to nurses in hospitals throughout the Commonwealth of Virginia, and we formalized a valid statewide surveillance program – the first in the country. The program was funded by the Centers for Disease Control for four years and subsequently by the Virginia State Department of Health.

Within three years, we were able to mine the data from over 1 million admissions to 85 hospitals in the State, and we were able to estimate that 5% of patients entering hospitals in the Commonwealth of Virginia acquired a nosocomial infection. When I subsequently moved to Iowa, I sought opportunities to develop a national, not just regional database. The SCOPE program (Surveillance and Control of Pathogens of Epidemiologic importance) was initially funded by Lederle laboratories, and with the strong support of two microbiologists at Iowa, my former fellow and now colleague, Mike Edmond and I enrolled hospitals nationally. It is now funded by several pharmaceutical companies and directed from Richmond, Virginia. The network has approximately 50 hospitals across the country under a continual prospective survey for nosocomial bloodstream infections, and we now have data on over 25,000 nosocomial bloodstream infections in the U.S. I have always believed in the need for active and continual surveillance to determine the causes and rates of serious infections. We can track rates of bloodstream infections and the proportion of antibiotic resistant pathogens over time. We also know the outcome of these infections, whether patients lived or died. With this system we know that *S. aureus* and enterococci are leading causes, and approximately 25% of the enterococci are resistant to vancomycin, until recently the last antibiotic known to be effective.

One of the behavioral issues that plague hospitals and especially the infection control team is how to improve handwashing compliance, how to achieve a plateau above the usual threshold of 40% observed in modern ICUs. This is not a new story.

In the 1840s, a young upstart by the name of Dr. Ignaz Semmelweis moved from his home in Budapest to begin the practice of Obstetrics in Vienna, the center of excellence in Europe in the mid 19th century. The major hospital-acquired infection at the time was puerperal sepsis, a life-threatening pelvic infection of women occurring immediately after childbirth. The Latin *puer* is "a child" and *parere* means "to bring forth" (from which "parent" is derived). So "puerperal" means "to bring forth a child." We now know that this infection was due to group A streptococci. Group A streptococci and enterococci are distant cousins on the evolutionary tree of bacteriology, but the 1840s were 40 years before Pasteur's experiments in Paris elucidating the germ theory of infection.

It was a defining time at the general hospital where Semmelweis had contracted to work, because the relatively new medical practice, the gross autopsy, was being utilized coincident with the emergence of group A streptococci. For the first time physicians could routinely examine the diseased organs of patients who died and infer the cause of death by classifying the pathological changes. Because microscopic examination of tissue from organisms was not done initially, and only the slices of organs visible to the naked eye were examined, the name "gross" autopsy was used. By today's standards the gross autopsy seems a trivial advance, but in the 1840s it was the MRI (magnetic resonance imaging)- equivalent of medical technology. It soon became common practice for physicians and medical students to review the gross autopsy findings of women who had died of puerperal sepsis and subsequently to begin ward rounds, examining in succession the women who were in labor.

Semmelweis made a number of observations, having been the first to perform accurate surveillance. Compared to the 2% mortality for women hospitalized on alternative wards where midwives alone delivered the mothers, the mortality on wards attended by physicians and medical students was 8%. Formal statistics were not applied to data at this period, but the results showed obvious disparity in outcomes for two otherwise similar patient groups. This was disturbing news to the extremely authoritarian administrators of the hospital but recognized through local gossip by wary families of patients, who often delayed their visit to the hospital to the day when all new patients would be managed by midwives. Regrettably, Semmelweis was not a man of tact, and in the bright lexicon of rigid idealists, he pointed an accusing finger at his colleagues and called them "killers." His lack of political tact would not be forgotten.

Semmelweis also noted that the mortality rates improved whenever the medical students went on vacation, and that the smell of the autopsy room was present whenever the students were on the wards. Importantly, one of the pathologists by the name of Kolechka accidentally nicked his finger with a scalpel while performing a post mortem examination on one of the women who had died of puerperal sepsis. Not only was Kolechka's clinical course similar to that of women dying from puerperal sepsis, but also when he died, the gross autopsy findings - the MRI equivalent - were remarkably similar.

Reasoning that something must have been carried from the autopsy room to the women in labor on the hands of physicians and medical students, Semmelweis introduced chloride of lime handwashing solutions to be used between patient contacts. As a result of this thoughtful intervention, the infection rates and mortality were documented to have fallen, the latter to below 2%. Follow up surveillance data illustrated the benefits of the intervention – handwashing.

In the minds of his superiors, still bitter from his harsh accusations, Semmelweis was not, however, a hero. Despite his remarkable talents and accomplishments, he was relieved of his post and forced to return to Budapest, eventually encountering similarly high rates of death among young women in labor and again showing correspondingly favorable results with the introduction of hand cleansing.

In his lifetime Semmelweis would never receive honor, but modern infection control experts regale his courage, his insights to perform surveillance, his hypothesis regarding hand carriage of a harmful substance decades before the germ theory, and his implementations of successful life-saving interventions for control.

Semmelweis promoted facts over bias through active surveillance. He studied populations and relied on data instead of individual anecdotes. And he was a man of principle, not politics when it came to saving the lives of young women. His greatness, therefore, is not the product of his intellect but the result of his personal commitment, his discipline – and in a phrase, his moral force and high standards.

I have been inspired by the data of Semmelweis and strongly encourage handwashing in the hospital. I found out that patients themselves can be a powerful social influence in handwashing compliance by the health care team. Earlier in my career I had seen a 31-year-old woman who was admitted to my service one evening when I was on call at UVA. She had had high fever and chills, a rapid pulse rate, an elevated breathing rate, and an extremely depressed white blood cell count. What was especially unsettling was the fact that many of her white blood cells when examined under the microscope were immature. The question raised by the laboratory technologists was whether she in fact had acute leukemia. She herself described an acute illness only, and in the weeks preceding this illness she had no symptoms of weakness, fatigue or weight loss that might suggest a more chronic underlying process.

We drew our blood cultures and treated her broadly for an anticipated *Pseudomonas aeruginosa* bloodstream infection, because it is the most difficult organism to treat effectively in the company of acute leukemia among patients with depressed white cell counts. I consulted one of the oncologists on call that night to give an opinion, and he suggested that we begin chemotherapy right away for acute leukemia.

Sometimes infection can cause a slight shift towards immature white cells with ten percent or more "bands," the immediate precursor to the mature neutrophil. This is a type of recruitment in which the infection signals the bone marrow precurser cells to assist the mature neutrophil to help out. As a result, the marrow pushes out the somewhat immature band forms. But our patient had cells representing two generations before the bands, extremely immature myelocytes and metamyelocytes. For reasons that are unclear, the recruitment of very immature forms of white cells occurs in some infections.

Quite anxious about the case, I declined the advice to start chemotherapy, decided to wait until the next day and follow the patient's course carefully overnight. I also wanted an opinion from a trusted friend who was another hematologist-oncologist, especially because the side effects of chemotherapy can be severe and disabling. In fact, I was not convinced of the diagnosis of leukemia.

The patient was fully alert, had a strong personality, and understood that very low white counts of any cause represented a risk of still more severe infections, the types that can be acquired in the hospital. I explained that handwashing was very important. For the most part organisms go from person to person in the hospital on hands, and increasingly these are resistant to many antibiotics. Occasionally a few organisms like Strep are transmitted by large droplets from the throat of one person to another after a cough or sneeze or even just talking. Still less commonly are true airborne infections in which a person can cough, leave the room, and hours later the organisms can infect someone breathing the air.

I asked my patient that if she was comfortable with an unconventional idea: I would ask her to request, really demand, of any healthcare person – nurse, physician, student, lab tech and others – to wash his or her hands before examining her or touching her IV line. She was a willing participant in her health care, and with conviction and respect made this request, saying that "Dr. Wenzel wants anyone who approaches me or my I.V. line to wash their hands first."

I soon learned that this was not only effective but also a socially powerful influence. One of the medical students recorded his interaction with the patient in some detail on the chart, perhaps after an embarrassing encounter with my patient. Subsequently, on many occasions I have asked other patients who were confident that this would not jeopardize their relationship with the medical staff, to do the same. In each case the patient would say that it was my idea – not theirs. I was always prepared to defend their statement – and their well being. If they were timid in any way, I would not suggest that they participate in the social experiment.

The next day my colleague in oncology agreed that it was premature to conclude that my patient had leukemia, and we awaited the results of the bone marrow aspirate while I treated for suspected bloodstream infection. After 36 hours, both the urine and blood cultures yielded *Proteus mirabilis*, a much less aggressive Gram negative rod than *Pseudomonas aeruginosa*. The odds that the patient had an underlying leukemia were markedly reduced with the finding of an association with Proteus.

The organism is so named because when it grows on agar, it often does not form discreet colonies but instead swarms across the agar like waves. Its unusual form and name refers to the old sage and prophet of Greek mythology who knew all things - past, present and future. Because Proteus would not yield this gift of knowledge willingly, he would have to be chained while fast asleep by those seeking his secrets. In order to avoid capture, however, he was known to have the ability to change himself to various forms such as a lion, a dragon or wild boar. Thus, the appellation of Proteus for the Gram negative rod is in recognition of its ability to assume an unusual colonial form on growth media.

The patient's white cell count began to budge upwards to a safer range. Over a period of five days it would slowly return to normal, and fortunately the bone marrow aspirate would rule out leukemia. In fact, what brought this woman to the hospital was a complicated urinary tract infection: she was one of the three percent of patients who develop a serious secondary bloodstream infection – a syndrome we call urosepsis.

The body's response to this Gram negative rod was to throw everything it had in terms of scavenging white blood cells at the pathogen in the face of its coincident suppression of the usual rise in mature neutrophils, the trained soldiers on the front line. It's as though in a panic the body sent immature recruits and untrained militia, the non-mature cell forms of the bone marrow, into the bloodstream, because as a result of the infection there was a transient failure to muster the mature forms.

While my patient was recovering from her original infection, she fortunately did not acquire a hospital acquired infection as a result of her low white cell count. Possibly excellent handwashing by the medical team was important. For years I have recommended enthusiastically that all patients respectfully employ the lessons I learned about social pressures in medical care. There is nothing impolite or arrogant about asking every physician, every nurse, and every technologist to wash his or her hands before touching a patient or any device attached to a patient. It takes only seconds but can provide years of extended life by averting a serious infection. It is the right thing to do whether viewed socially, culturally, economically, medically or ethically. I will always be grateful for the courage of a 31-year-old woman with a Gram negative rod in her blood who led me to this conclusion. Thus, as part of an infection control program while working on any outbreak, I offer this approach whenever patients are willing and comfortable. When performing follow up surveillance to verify to effectiveness of control measures, it is not possible to ascribe a particular intervention with a measured proportion of the success. However, I am convinced that it helped in the control of epidemics caused by a single clone of bacteria.

One hundred fifty years after Semmelweis, we still have problems in achieving excellent handwashing compliance by health care workers. It is not entirely clear why very strict handwashing is not observed. I have a few ideas, however. Many physicians and nurses say they are too busy, that urgent situations force them to move from patient to patient without the time for handwashing. I think that this occurs, but only infrequently justifies a lack of handwashing. Instead, I think that health care workers trivialize this basic issue perhaps recognizing the fact that nine times out of ten, the absence of handwashing leads to no infection. The tenth time it may, but the link between a single episode of no handwashing and a subsequent infection is very vague. It is not that medical workers are unaware of the germ theory but rather what occurs is some repression of their possible role in transmission of infections. And for the most part there are not consequences to the physician or nurse for no handwashing: few supervisors make it an issue; the hospital does not dismiss non-handwashers; the joint commission reviewing the hospitals' credentials does not make it a big issue.

At the University of Iowa my team was the first to have had the opportunity to compare alternative handwashing agents and measure their effects on rates of infection among patients in three different critical care units. The standard agent was chlorhexidine, a medicated soap available in many formulations, in part because it was usually safe to the skin and importantly because it had excellent effects at removing or killing Gram positive cocci such as *Staph aureus*, group A streptococci, and the Enterococcus. The alternative was an alcohol solution sold with an emollient to prevent drying of the skin and dermatitis from frequent use. Alcohol has been the preferred agent in some countries in Europe, such as Finland and Germany, because it is very rapidly acting and inexpensive. Interestingly, the prevailing thinking in these European countries is that alcohol is milder to the skin of the hands than is chlorhexidine. In contrast, in the 1990s most nurses in the U.S. thought just the opposite, that alcohol was drying to the skin and more harmful to use repeatedly.

We had a large team led by two fellows, Doctors Gail Stanley and Brad Doebbeling, and supported by a wonderful group of study nurses. After several months' education period, we performed a study lasting over 8 months that alternated the hand cleansing options on a monthly basis in each of the three ICU's. Over 1600 patients were observed in each study arm, and chlorhexidine appeared significantly better than alcohol at reducing the rates of nosocomial infections. I was disappointed because I thought that alcohol was simpler to use and less expensive than chlorhexidine. I thought that we would show an advance in handwashing in hospitals. On closer examination, however, nurses and physicians used a much greater volume of chlorhexidine than alcohol with each handwash, in large part because of the prevalent belief among nurses that alcohol was too drying for the skin.

For the study we measured not only the rates of infection, we also measured the rates of handwashing compliance and found it to average 40%, implying that 60% of the time when handwashing was warranted, it was not carried out. When handwashing compliance was examined within medical disciplines, the rates were better among nursing staff than among physicians. In fact, similar rates of handwashing had been reported in journal articles from other university hospitals, and none of my counterparts at other institutions were surprised by the data. Our abstract was presented at a national meeting, and our paper was accepted for publication in *The New England Journal of Medicine*.

The national press picked up the story, but several articles focused on the 40% average handwashing compliance, a fact not lost on one of the senior hospital administrators at Iowa. I was invited to a meeting attended by administrators and a small number of colleagues from different departments and essentially scolded for airing out the dirty laundry. The disgruntled senior administrator implied that the reporting of handwashing compliance was not warranted in a high profile publication. I was immediately uncomfortable for many reasons. It was the first time that I experienced what I thought was a thinly veiled attempt to censor scientific publications by an academic hospital administrator. Furthermore, despite my full commitment to the control of infections in the hospital, was I the target of attempted intimidation by a hospital administrator who suggested that my actions were not supportive of the hospital? One can't help to recall the fate of Semmelweis at these types of meetings.

Fortunately, after approximately 30 minutes of hand wringing, the administrator opened up the discussion, and I was elated when an Associate Professor in the Department of Pediatrics announced indignantly and with strong emotion that an academic center was surely not a place for censorship. He was both a scientist and a scholar and wanted no part of it.

That statement turned the emotional tide of the group in the meeting but not the heart of the administrator. Nevertheless, the meeting ended with no recommendations. However, as the principal investigator of the study, I had spent some political capital at my hospital with the report that remains in the craw of a few hospital administrators years later. Fortunately, our manuscript and all details of the handwashing compliance were published with no censorship of the data. It is an important paper, and the vast majority of readers focused on the study design and key findings that one agent was preferred over the other, leading both to better handwashing compliance and a reduced rate of infections acquired in ICUs. All of our team were proud that we made an important contribution in the literature. Although I may be proud of being in the great shadow of Semmelweis, I nevertheless warn all young fellows in training that a few hospital administrators similarly may have professional and emotional links to their corresponding role models in the Vienna of the 1840s.

My own experiences trying to control ICU-related infections also offered a lens by which to view the ageless force of politics in the academic center. The fact is that medical and administrative disciplines do not always share the same values. Most importantly, there is a harsh reality that at times the truth can hurt. However, there needs to be no conflict between scientific integrity and institutional loyalty. The key I think is to focus on the patients and the Hippocratic wisdom – "above all do no harm".

Improving handwashing compliance to reduce the rates of serious hospital-acquired infections remains of great interest to me. In 1997, I would have the opportunity to study the acceptance of a new and increasingly accessible alcohol-based hand cleansing agent compared with the traditional sink/soap combination. By the end of the 1990's it should be pointed out that a major social view of alcohol hand cleansing agents had gripped the United States. Increasing numbers of boutique stores carried small containers of alcohol preparations of various colors and fragrances. Women carried alcohol hand cleansing tissues or liquid in their purses, in their cars, and had them available at home. The cleaning of hands with alcohol had become widely accepted socially as a means of reducing infections in the community.

We sought the opportunity to restudy alcohol as a hand cleansing agent in the hospital. Our question was whether compliance would be increased. Leading our study was a young German physician from Hanover, Werner Bishoff, who wanted to spend a year learning infection control with our team. Working day to day with our full time study nurse, Werner directly observed handwashing compliance in our Medical ICUs: before and after an intense educational program, after making an alcohol-based hand cleansing agent available to the ICU team – 1 for every 4 ICU beds, and then after making the alcohol product available in front of every ICU bed. The alcohol preparation is very convenient: with general care of patients, a nurse or physician could cleanse hands quickly with the alcohol product, which requires no queuing in front of a sink, no use of soap or water, and no use of towels, because it dries quickly after hand rubbing.

Baseline handwashing compliance was poor, 9% before and 22% after touching individual patients – at best 31% total. After intense education, compliance changed minimally to 14% before and 25% after – at best 39% overall. However, after introducing the new, increasingly accessible alcohol-based ("waterless") hand antiseptic, there was a statistically significant ($p<0.05$) improved compliance: 60% with one dispenser available for every four beds. It was even higher with one alcohol dispenser per bed: 23% before and 48% after patient contact - total 71%!

I conclude that more education is not what is needed to encourage health care workers to wash their hands. They already know it. The answer is to make the process simple. The answer to compliance is simple: access, access and access! No sink dispenser, no soap, no queues for handwashing, and no towels are needed. We have now made alcohol hand cleansing dispensers available throughout the Medical College of Virginia Hospital. In April 2000, our study was published in the *Archives of Internal Medicine.*

I have identified a number of epidemics of nosocomial infections, many related to antibiotic-resistant bacteria. Finding a rate higher than expected was only possible by having a clear understanding of expected rates, and this was possible only with an active surveillance program. Furthermore, when we introduced a specific intervention as a means designed to control infections, we can only know its value by seeing reductions in the background rate.

Many hospitals abandoned surveillance programs in the early 1990s as cost control efforts, and the assault of managed care forced institutions to reduce budgets for infection control. Furthermore, surveys that need to be performed week after week can wear on any professional. Some in our field conclude that surveillance is not needed for infection control today. As a result, I am deeply concerned that more and more U.S. hospitals today know less and less about the rates of infection in their respective institutions.

There may be infection control without surveillance, but those who practice without measurement, practice without one of the major tools of science. They will be like the crew of an orbiting ship traveling through space without instruments, thus unable to identify their current bearings, the probability of hazards, their direction or rate of travel. Only their indifference provides temporary calm. With such disorientation and oblivion, they should be terrified.

THE ORIGIN OF FEAR
(Neisseria meningitidis)

While entertaining friends at dinner one evening in Iowa City, I received an alarming call from our friend and neighbor Milt Rosenbaum, and he was greatly concerned about his wife's health. Milt and his wife Irene had both worked at the University of Iowa. Milt, a renowned professor of Psychology, was still famous for his marching with student protestors in the mid 1960's to protect them from any possible harm from the Iowa state police. Kent State was then a close memory, and Milt shared his liberal feelings with students in a paternal fashion. The police chief was in fact quite wise and suggested that students wishing to be arrested kindly walk to the bus. Milt encouraged a non-confrontational protest, and no one was hurt. He was fearless in his pursuit of non-violent protests of the government and rejected both racial prejudice and the Viet Nam War.

Irene had been a social worker helping families of patients on kidney dialysis. She played golf with Milt and fully supported his iconoclastic and humorous view of life. They had both grown up in the Bronx and forever retained their love of New York and modern jazz. They both loved world travel, and Milt especially was enamored with France, the wines of Provence and characteristically very rare, broiled lamb chops served with a French Bernaise sauce. They had many friends in art, history and psychology, and our lives at Iowa City were enriched by a friendship with the Rosenbaums.

Milt and Irene had just returned from a visit with their daughters in California. Irene had begun to feel uncomfortable on the last leg of the flight – perhaps just simple exhaustion from travel they thought, but she became so lethargic that she went to bed before dinner. Milt had just gone upstairs to check on her, and he was unable to arouse her.

After fetching my stethoscope, I went immediately to examine Irene accompanied by our dinner guest, Mary Nettleman, a physician and friend whom I had recruited from Indiana University to my division. When we arrived at the upstairs bedroom, Irene was indeed comatose. Even pressure on the sternum with my knuckles, that would usually be a painful stimulus, failed to arouse her. On touching her skin, I felt great heat. No thermometer was needed to confirm that she had a high fever. A quick examination of her eyes showed that her pupils were equal and normal in size. Had they been unequal she might have been suffering from a stroke. But a high fever with stroke would be unusual. Her neck was supple, no sign of meningitis. Her lungs, heart and abdomen were normal, as were her neurological reflexes.

What was abnormal was seen when I pulled down the lower lids of her eyes and examined the pink conjunctivae. There was a single petechia, a tiny bleeding spot, on the conjunctiva of each eye. She had sudden coma, fever, and rupture of tiny capillaries. She probably had disseminated meningococcal disease.

I dialed 9-1-1 and within minutes the police, ambulance and fire engine arrived at our quiet neighborhood. The paramedics transported Irene to the Emergency Room, and she was admitted quickly to the ICU. A stain of spinal fluid did show Gram negative diplococci, and subsequent cultures of both blood and cerebrospinal fluid would yield *Neisseria meningitidis*.

Irene's breathing would be supported with a respirator for four days, but she emerged from a coma after 48 hours. Nevertheless, she later told me that she remained lethargic and weak for eight months afterwards. Irene eventually returned to full health, but if Milt had left her undisturbed through the night, she would have died in her sleep, a victim of this terrifying organism, another entry in U.S. vital statistics data within hours. Irene had gone from feeling fatigued to a deep coma with fever and blood clotting abnormalities. Although she had bacteria in her spinal fluid, her neck was not stiff because we saw her so early in her course, hours before inflammation had time to set in. With antibiotics, great critical care support at the University of Iowa, and luck she made it.

Fortunately, experience leads to expertise. In my training, military service and career as a faculty member, I learned to consider and recognize this uncommon infection. Such experience is important and probably could be quantified. For example, an analogy in surgery is the well-known relationship between high volume and best outcomes. If you need an operation and have three surgeons available from which to choose, a single question will probably lead to the best choice, all other things being equal: "How many do you do each month?" For an infectious diseases specialist, the question translates to, "how often have you managed such cases?" I had consulted on several. I was especially pleased to be helpful to the Rosenbaums.

One year earlier a particular incident had endeared me to Milt. On an unusually warm August evening, we were having dinner with close friends from our Charlottesville days. Visiting from Baltimore were Mike and Trina Johns. At the time Mike, an ear nose and throat cancer surgeon, was Dean of the School of Medicine at Johns Hopkins. Trina thought that she saw a bird flying in the house, and within minutes Mike and I put on proper dress to capture the bat that had invaded our conversation: long coats, bike helmets to avoid having the beast land in our hair, leather gloves stored from the winter, and tennis rackets.

We managed to frighten the animal into the direction of our back room where we usually watched TV and from which a doorway led to the outside. Immediately we sealed egress from both inside entryways, isolating the bat to the war room. Mike and I called for my wife, JoGail, to open the outside door freeing the bat to leave the premises. Seeing both of us cower beneath ridiculous outfits, she simply inquired, "What is it about this situation that makes me immune from rabies and you two physicians at high risk?" She opened the door to the outside, and the rest of us were hoping and expecting the winged mammal to flee for its life.

We assured JoGail that the bat had flown swiftly through the door but just needed her careful confirmation. No sooner had she reentered the TV room than the flutter of wings in flight announced how wrong we were. JoGail tried to chase the bat with the tennis racquet, but the 9-foot ceilings offered too much free space.

In the previous year we recalled that over a period of about four weeks, Mitt had fearlessly trapped over eight raccoons that had invaded his garage, and then one by one he drove the furry prizes to a remote site ten miles away for their eventual release. Because an easy capture of our bat was elusive, we called Milt.

Arriving in short pants, a golf shirt, and sandals, Milt Rosenbaum was in his early sixties, very thin but quietly confident. His curly gray hair piled high in erratic patterns on his head gave him a professorial appearance. Milt was smoking a cigarette, and he carried a green fishing net with a 2-foot handle. Briefly inspecting Mike and me who were still sporting our bat hunting gear, he seemed incredulous but entered battle mercifully without a deprecating word.

Within seconds Milt returned with a bat fully entangled in the mesh of the fishing net. He would soon release it outside without incident and with recently won pride. On leaving the kitchen at that moment, he had a swagger like John Wayne, his cigarette dangling from the corner of his mouth like James Dean, and he was silent in his heroic deed. Though he uttered nothing as he walked proudly into the sunset, he glanced back briefly, and somehow I heard the words, "Chicken Shit!"

People without irrational fears of bats may wonder how a person like me with 25 years of formal education – 12 years of primary and secondary school, 4 years of college, 4 years of medical school, and 5 years of subsequent training – could react the way I did. In contrast, my neighbor never hesitated to respond to our crisis. The crises in medicine stand in stark contrast, however, since decision making is increasingly based on knowledge, relative benefits and side effects of options, experience, wisdom and ingenuity. Phobias rarely are an issue, and if we as physicians stay focused and continue to improve, we become successful.

If someone were to ask the question, "how do you know if a person is successful at his/her position?" the answer might be that he or she has learned sufficiently to make good decisions. In medicine one can learn from experiences with diseases involving either individual patients or a population. A clinician builds a body of knowledge from one-on-one encounters with a series of patients. An epidemiologist or a public health authority learns from his experiences in developing views of health policy for a population.

I do not claim to have made good decisions always, but the meningococcus, a single organism, on more than one occasion offered me opportunities to test myself in both the clinical and health policy arenas. Experiences with individual patients gave me important clinical skills, and with two small outbreaks brought home some important lessons with respect to populations.

One night while on the infectious diseases service at the University of Iowa, I was asked to see a young man in his 20s with meningitis. Our on-call team was told that he had probable pneumococci on Gram strain, Gram positive diplococci. As faculty members we always stressed that it was a good idea to examine the stain ourselves, and I went with the fellow and medical student to the microbiology laboratory, placing the slide with the smear from the cerebrospinal fluid under the lens of the microscope.

To my surprise we saw not Gram positive but Gram negative diplococci. This error can result from faulty technique in applying the stains, problems with the stains themselves, and problems with interpretation if enough of the fields of the bacteria are not viewed. The results, however, indicated that the patient had meningococcal, not pneumococcal meningitis. The importance of this finding is not that therapy changes. Fortunately, at that time both bacteria were susceptible to penicillin and to a cephalosporin called ceftriaxone, two alternative antibiotics. What does change is managing contacts and the use of isolation procedures, because meningococcus can spread to nearby patient contacts, even if infrequently. However, illnesses from pneumococci do not result from person to person contact in a hospital setting, prophylactic antibiotics are not needed after exposure to patients with pneumococcal infection, and no mask would be required when treating patients with this organism.

Our young patient would do well with antibiotic therapy, but an additional consideration was that he was a student at the University of Iowa who had just returned from spring vacation after a 30 hour chartered bus ride with 51 fellow students. An index case with infection could possibly transmit the organism to others during the long ride in a confined space. We would have to identify and find all contacts and treat them prophylactically with an antibiotic to prevent serious infection.

Meningococcal species can live in the mucous membranes of the pharynx, and most of us will eventually be colonized for weeks or months with a benign strain and eventually develop antibodies to all sister strains, including those aggressive ones capable of causing serious disease. However, if close contact allows virulent organisms to travel from a carrier with protective antibody to a "virgin" with no prior experience with meningococcus, serious disease can result. Secondary cases of disease from an index case are uncommon, but occasionally travel from a patient to a family member or to a roommate in close contact, essentially by exchanging saliva – perhaps by sharing a drink from the same glass or bottle.

With the help of some students on the bus and the doggedness of the infection control team – called into action at about 9:30 p.m. – we located all but six student contacts of the patient by midnight. The remaining members were located by noon the following day, and no secondary cases occurred. The clinical point is that physicians need to be directly involved in reviewing the laboratory findings themselves. Errors occasionally occur, and our personal review of the Gram stain may have prevented a misdiagnosis and mismanagement of contacts. In fact, the odds were low that anyone would have acquired disease without prophylaxis, but the risk was not zero. If we prevented a single case, it was all related to a simple staining procedure designed by Danish botanist and physician, Christian Gram, 100 years earlier in the mid 1880s.

Even before arriving in Iowa I had great respect for the awesome meningococcus. They mingle freely among the society of less virulent pathogens and initially go unrecognized. Subsequently they strike quickly like evil tyrants, killing one person visibly and violently in a community, and instantaneously unleashing fear in the hearts of thousands nearby. Close friends and family members become horrified because only hours earlier it seemed that the dying person was vibrant and full of hope. Only a curious physician with experience can readily identify meningococcal disease in its early stages, and I admire the primary care physicians who kept this devastating illness in mind while managing ten to twenty cases of influenza each day, a relatively benign infection with similar clinical presentations to the meningococcus. Yet after infection with *Neisseria meningitidis* – the meningococcus – patients can progress from feeling mildly ill to being moribund within hours of the onset of bloodstream infection or spinal meningitis, a serious infection of the thin layer of tissue surrounding the spinal cord and brain. Mostly, infants, children, and young adults are affected, but it sporadically infects people of all ages. It all seems so random and unpredictable.

When stricken, patients with meningitis become profoundly lethargic, develop high fever, and often say that they have the worst headache of their lives. Within hours, the neck loses all flexibility, the rigidity of the muscles developing as a protective response of the body to the inflamed meninges. One or two pink colored spots appear initially on the chest, arms or hands, then spread to the rest of the body. The pale red discoloration gives way to a purple appearance, the small lakes of bleeding in the skin. The infection causes havoc with the usual cascade of clotting factors in the body, inciting simultaneous bleeding and clogging of the small arterioles. Pinpoint bleeding spots called petechiae appear early or late, telegraphing the bursting of tiny capillaries. Despite the bleeding, the blood supply to vital organs is relentlessly choked off by clotting. In the end the patient has labored breathing, is overcome with dread, and continues to bleed. There may be gangrene of the fingers and toes announced by the coolness of the dusky colored digits. The heart rhythms become erratic causing the patient to sense palpitations, and the lungs fill relentlessly with fluid choking off the air supply. The kidneys go on strike refusing to filter the blood appropriately and to discard the body's toxins in the urine. When the brain continues to swell within the immutable confines of the bony cranium, coma ensues and death arrives quickly.

Penicillin therapy administered early is life saving, and in the year 2005, there were excellent treatment options including broad-spectrum drugs in the cephalosporin class. In severe cases, ICU care is essential for respiratory support and cardiac monitoring. The key point is to consider the diagnosis early and to treat as quickly as possible in order to save the life of the infected victim. We clinicians attempt to maintain a high level of alertness by presenting unknown cases to each other at a weekly case conference.

At the University of Iowa all cases of meningococcal meningitis seen in the institution would be presented at "ID Rounds." Pertinent clues from the case histories and physical examinations were presented by fellows in infectious diseases to the gathered assembly of faculty. The cases were presented as unknowns, and the faculty were asked pointed questions by the fellows in the attempt to "stump the stars." At times the fellows would withhold very important information. We faculty would of course ask for more data, seeking a critical fact, stalling for time to think and hoping to say something important to ward off an appearance of ignorance which might be humiliating. Some of the fellows were quite clever and learned the art of the not so gentle but acceptable putdown: "All right, Dr. Smith, that was state-of-the-Art in 1975, but how would a modern specialist treat the disease?"

One of the key questions might easily be, "What risk factors in the patient made him susceptible to the infection?" The answer includes a lack of specific protective antibody to the meningococcus. Obviously close contact with a carrier of a virulent stain is important, but patients would rarely ever know this. It is likely that some genetic predisposition exists because a clinician asks, "Why if five people were exposed to a carrier and all five had no protective antibody, did this particular patient become ill at this time?"

Occasionally small clusters of infected patients or epidemics of meningococcal disease add a new layer of complexity. In such situations, we not only have to treat individual patients but also identify the cause of the epidemic and control it. In the early 1990's, we saw two students from the University of Iowa over a two-week period with serious meningococcal disease. One was a 20-year-old young man with meningococcal meningitis and bloodstream infection. The other was an 18-year-old woman with both bloodstream infection and a shoulder infection from meningococcus that had found extravascular lodging within the cartilaginous cushion of the joint. As the hospital epidemiologist I was asked to investigate these cases and offer advice to University President Hunter Rawlings.

To identify similarity or clonality of isolates, it is necessary to do some genetic fingerprinting of the organisms. For a number of years the practice was to grow the organisms, extract their DNA or genetic code and "cut" the DNA into pieces with enzymes that choose specific locations along the DNA. Then a technologist would expose the cut pieces of DNA, that are embedded in agar, to an electrical current. Depending on the size of the fragment and its charge, each piece will migrate at different speeds and thus arrive at varying distances from the starting line. The result is a ladder of lines representing the small fragments of DNA that moved various millimeters from the initial well. Identical strains will show identical migration of the fragments with the irregularly placed "rungs" of the ladder in exactly matching positions.

When bacterial fingerprinting of the meningococci from the two students was done, it was obvious that both victims had been infected with the identical clone of bacteria. The laboratory data showed that we had a mini epidemic, an "n" of two. The question in an epidemic situation is not what risk factors make a patient vulnerable, but what risk factors make populations vulnerable. What prior exposures did they have in common?

In fact, they were not roommates or friends or classmates. They didn't even know each other.

Both students had attended the same collegiate football game one week prior to the illness in the first case. The contest was played on the campus of the University of Illinois at Champaign Urbana, an important fact, because the University of Illinois had experienced an endemic or smoldering problem with meningococcal disease among students between February 1991 and April 1992. There were nine infections and three deaths, and in February 1992, 16,000 students at the Illinois Campus received meningococcal vaccine.

At Iowa, the local public health personnel, student health physicians, and the hospital epidemiology group met to discuss the cases. The student health team was led by Mary Kowassah, M.D., an excellent administrator with good contacts across big ten colleges.

Mike Edmond was the young fellow in training with me then, and Mike led the investigation with a meticulous style that was his characteristic approach to epidemics. He managed to get a sample of isolates from our two cases of meningococcal infection in students as well as some from the cases in Illinois. With help from one of our microbiologists, Mike Pfaller, we did our own fingerprinting at Iowa and found that the Illinois clone and the recent Iowa clone were identical. The implications were straight forward: Sometime during the weekend surrounding the Iowa-Illinois football game, our University students had acquired the meningococcal organisms, probably from students at Illinois who were asymptomatic carriers. Transmission was unlikely to have occurred outdoors at the football game itself because it takes intimate contact to transmit meningococci. We thought that it was more likely to have occurred indoors, perhaps in the tight quarters of a fraternity party. Colonized students from Illinois had in some fashion exchanged saliva with our undergraduates, neither of whom had had protective antibody to the bacterium. Whatever the mechanism, the Illinois strain of *Neisseria meningitidis* had arrived on our campus.

What was initially confusing and worrisome was the fact that three more cases occurred in Iowa City between November 19th and December 15th, 1992. One patient was a 22 year old bartender, the second a 21 year old male student, and the third a 21 year old woman who was not a student. Of interest, *none* had attended the Iowa-Illinois football game, and they were not even friends of the initial two undergraduates or of each other. However, the second and third case had frequented the bar where the first case was a bartender.

When we did fingerprinting of the DNA of the five Iowa strains, the two students with the Illinois strain were totally different from the other three. However, the three species with common exposure to the same bar were the same and were obviously linked together. Even before our publication, we shared our findings informally with other collegiate student health services. We obviously had two small epidemics of meningococcal infection in our community, distinguished only by modern molecular biology techniques.

University President Hunter Rawlings, his provost, and the Dean of students asked for advice. Of course we had given rifampin to close contacts of each of the patients. This drug accumulates in high concentrations in the oropharynx – unlike penicillin which works only if there is inflammation as is seen commonly with infection from a "strep throat." Rifampin is an excellent option for eradicating the carrier state, killing any newly acquired meningococcus residing in the throat. But the only long-term hope of preventing continued cases among University students was to give meningococcal vaccine. We had hoped to learn from the unfortunate experience at another university in developing new policies for the health of students.

The vaccine would cost approximately $10/injection, and if we vaccined 20,000 or 25,000 of the student body, that would amount to at least $200,000 or $250,000 dollars. "Could I guarantee that this would prevent new cases?"

"No," I said, "but the probability is high." I reframed the issue, "Can we afford another case or a death in the student body if we fail to vaccinate?"

In fact, Hunter never hesitated and was genuinely worried about the students, not out of fear of adverse public relations but of his interest in their health. For me the entire discussion was novel. Up to that point, I had never recommended a decision that had cost so much and had it accepted so quickly. Within an hour, a quarter million dollars were earmarked for vaccination. As a result I was quite anxious and hoped that we had seen our last cases.

Mary Kowassah organized a team of vaccinators for what became a "happening" at the University. Over 3,500 students every day for five days received meningococcal vaccination following carefully worded letters to parents and students, repeated notices about vaccination times, and visits to all dormitories and fraternities with strong admonishments for addressing the problem. In the two years of follow-up, not a single case of meningococcal disease of the epidemic strain was seen. An isolated case of group B meningococcal disease was seen in April of 1993, a strain causing sporadic disease but not covered by the available vaccine.

By the year 2005, more and more Universities were endorsing the use of meningococcal vaccine for all students. Previous decades of reluctance because of perceived poor cost-benefit ratios have paled in the face of increased press and scientific reports of tragic losses of life.

Periodically I return to a question I ask of myself, how does one learn in medicine? Specifically, what avenues are open for the acquisitions of new information and new perspectives? The answer is that a physician learns in one or more of three pursuits: one-on-one with patients – the clinical arena; with efforts in a laboratory to study the mechanisms of science; and also through a study of populations, using primarily the discipline of epidemiology. All are important, and each contributes a useful body of knowledge. Medical students need to travel down each avenue, and a medical school experience that leaves any one out of the curriculum will be wanting.

My own experiences with the meningococcus offered me the perspectives of individual patients, even an infection of a close friend, as well as the need to understand the disease and preventive measures within a population. Clinical and epidemiological skills are different and complementary, and the laboratory support that eventually grows out of bench research is an essential component of excellent medicine. The meningococcus has much to teach us from any perspective we choose.

A doctor spends most of his career constantly making decisions. Sometimes these involve life and death and have to be made rapidly. Usually the rapid decisions need to occur in the grave context of uncertainty, and this is the heart of medicine.

We wish we had more information, not only for the sake of the patient but also for our own comfort of mind. We seek assurance but rarely achieve it; and each decision has a boundary of time. The unforgiving minutes march on, and there is a limit to our period of consideration before a good decision must take place. Delays beyond that limit may have terrible consequences for the patient. But there is a second uncertainty: how much time before it gets critical to make a decision?

What I have learned about decision making is that experience is very helpful. If one has been in a similar situation before, early recognition of the illness and the importance of early and appropriate therapy will be of enormous comfort and value. Since we can't see every crisis, what else can we do? The answer is to imagine various illnesses in our differential diagnoses and reasonable responses, to be presented with challenging cases each week in an "unknown case" format and ask ourselves what we would do in that situation. This is the rationale behind the "Grand Rounds" that are blackboard cases presented in academic centers in every specialty. Another ingredient is to read, to keep up with the literature. And when we read about unusual situations to imagine how we would recognize and manage the patient. We physicians need continually to ask ourselves, "What if...?"

I have seen meningococcus on several occasions, and I keep it high on my list of treatable, rapidly life-threatening and contagious infections. Those who have never before had experience with it would surely think that if some patient looked like the worst case of flu they have ever seen, they might consider the diagnosis, perform necessary cultures, and give antibiotics. If they are right, a life may be saved. If wrong, they cope only with the side effects of antibiotics.

Similar considerations arise in managing epidemics of infection, but now cost and benefit enter the equation. Any society has limited resources, and some priority list needs to be compiled. One can see, however, how money and politics can influence policy for populations including the decision to vaccinate all college students to protect them from serious infections. Meningococcus has all of these lessons for us about decision making.

But training, conferences, reading the journals and years of experience – while very useful – will never be enough. There are many *new* situations, ones which we have never encountered. Our thinking at that time is this: given the situation, what are the likely causes, the most life-threatening concerns, the list of therapies that could manage all or most of these, and the safe therapeutic options? What is the cost of doing nothing, of doing some, most or all of the options I can consider? Often in the face of uncertainty we find that the patient or family members have the clues to a diagnosis. Physicians need to explore the details again until the new ideas become apparent.

A BUMP IN THE NIGHT
(*Bacteroides, Klebsiella pneumoniae, Plasmodium vivax, Coccidioides immitis*)

In seeking the cause of an obscure diagnosis, an effective clinician is obliged to obtain clues by asking patients the details of their social life. Often these details seem unimportant to an ill person, and a sick individual may have no suspicion that a recent trip to explore a cave was important in suggesting a pneumonia due to Histoplasmosis, a fungus that thrives in bat guano; or that the acquisition of a new pet salamander might have been the source of a gastrointestinal infection with Salmonella; or that a recent illness in the family's parakeet would suggest a pneumonia called psittacosis, caused by a Chlamydia species from the psittacine bird.

In working up a patient with a complicated disease, physicians begin with an understanding of the elements of the history of the present illness. Then we inquire in serial fashion about the details of the past history, the family history and finally the social history. With each successive step we are seeking to identify the characteristics of the current problem, any underlying problems and their therapies, and any genetically related illnesses. Lastly, the social history details the use of recreational drugs including alcohol intake, their pets, avocations, travel and leisure activities. All clinicians have to be medical detectives. All of this inquiry is completed before examining the patient.

Sometimes patients will offer the details spontaneously, and other times it requires repeated inquiries. I often think of the television character, Columbo, a homicide detective, who at the moment when he is about to end the series of probing questions of a suspect and depart, pauses at the doorway and apologetically says, "Excuse me, but I just thought of something and have one more question." Frequently this happens as physicians interview patients during the history and physical examination session. The issues of the social history can be especially important, and all too often ignored. I conclude, however, that all patients should have no anxieties about revealing these special aspects of their lives.

While at the University of Iowa, I examined a 58-year-old white man who recounted a curious event at a campsite 4 days earlier. He had an underlying diagnosis of adult-onset diabetes and was in the clinic for a routine follow-up. However, in answer to an open-ended question, "Is there anything new in your life?" he reported incidentally that while sitting by a campfire earlier in the week at about midnight he had suddenly and unexpectedly been "attacked by a large bird." Despite his stocky build and 220 pounds of weight, he was knocked from his director's chair by the force of the assault. Sometimes fears at night in the woods are justified and have nothing to do with panic.

My patient never identified the assailant but "saw feathers fly," and as an experienced woodsman was confident that he had been targeted by a large owl. In the middle of the night he had visited a hospital nearby the campsite where his scalp wounds had been attended, but no antibiotics had been prescribed. He pointed out to me and one of my students the two small lacerations on his bald scalp that were not red or swollen, and he had no fever.

A few days later, however, his scalp lesions were obviously red, and a small volume of pus was seen and cultured. I prescribed a quinolone antibiotic for broad coverage thinking that *S. aureus* or *S. epidermidis,* organisms likely to be residing on the surface of his scalp, might be the infecting agent or perhaps even a Gram negative rod. There are usually other therapeutic options, but he had serious allergies to both penicillins and sulfonamides.

Three days later my patient in a humorous mode showed up at the clinic wearing a baseball cap with an emblem that read "OWL FOOD," and now his scalp lesions were draining foul-smelling pus. All foul-smelling pus contains anaerobic bacteria, and at the bedside I suspected that we would find one or more members of anaerobes. On staining there were Gram negative rods, and the cultures of pus from his scalp wounds obtained in the previous visit were in fact now positive for anaerobes which eventually grew a *Bacteroides species*. These are organisms that live in the relatively oxygen-free environment of the gastrointestinal tract. Although the genus was identified, the species was too unusual to give a known name. It would never be identified. The wound was incised and drained, and a drug that targets anaerobes was prescribed, clindamycin. He was very much improved a week later.

I reviewed my case with a local expert on North American birds of prey including owls, members of the suborder *Striges*. It is not rare for great horned owls to attack noiselessly by virtue of their soft feathers. Apparently owls are attracted to white color, and when they have attacked skunks, they place their talons on the white part of the animal's fur preferentially. In fact, owls are one of the few creatures that attack skunks, apparently undeterred by their prey's obnoxious spray. I concluded that the white area of baldness illuminated by the campfire at midnight was the key risk factor for my patient. Possibly his underlying diabetes placed him at added risk for an infection after the razor-sharp talons ---- probably contaminated with the enteric organisms of a previously eviscerated animal --- tore the skin of his head.

I would later learn that the great horned owl, *Bubo virginianus*, can attain a height of two feet, a weight of 3 ½ pounds, and a wing span of up to five feet. They are aggressive raptors with few predators and are quite fearless. Their night vision is remarkable because their retina has a high proportion of light gathering rods. Because their eyes are not freely movable, the owl needs to turn its head to see elsewhere. Fortunately, its turning radius is a remarkable 270°! Their tufts or "horns" are not their ears, but the asymmetrically placed ears on either side of the head help triangulate the precise location of its prey. Subsequently, the great owl can eat its victim whole, later regurgitating up a pellet of undigestible elements. My patient had apparently been attacked by an amazingly evolved predator, a nocturnal specialist with extraordinary skills for midnight aggression.

In a brief note in a medical journal I cautioned that even with this entity – which I called "Striges Scalp: Bacteroides Infection After an Owl Attack" –it was premature to recommend wearing dark colored caps or head protection routinely in camp sites to ward off subsequent attacks by winged creatures of the dark. In subsequent teaching and lectures, I have tried to influence my students to ask details about hobbies, travel and pets whenever an unusual infection occurs in a patient. It helps to have a few stories for them to remember the lesson.

At the University of Iowa, I would add another story, having consulted on a 26-year-old man while on the infectious diseases service. Complaining of fever, he was admitted with obvious sepsis, and his blood cultures grew *Klebsiella pneumoniae*, a Gram negative rod that is frequently seen as a nosocomial pathogen or an unusual cause of pneumonia acquired in the community. This organism is named in honor of Theodore Albrecht Edwin Klebs (1834-1913), a contemporary of Pasteur and Koch, who was a native of Konigsberg where he studied Medicine. Particularly interested in pathology, Klebs presented his dissertation in Latin in 1856 on the effects of tuberculosis in the intestine. He lived for some time in Switzerland and made a key observation that the "highlanders," whose main food staple was milk and milk products, were susceptible to tuberculosis. Thus, before Koch discovered *M. tuberculosis* and related this specific organism to the pulmonary form of TB, Klebs as a result of careful observations showed that intestinal tuberculosis could be acquired orally by contaminated milk.

The young man on our wards was quite active and had no pneumonia or urinary tract infection. He had acquired his Klebsiella bloodstream infection in the community and not in the hospital, and in fact had no infection except the one in the bloodstream. Sometimes the mechanism of entrance of bloodstream pathogens remains elusive and only a careful review of the details of a patient's activities can illuminate the important facts.

The patient did extremely well on antibiotics, and a careful review of his past medical history revealed nothing. He, in fact, had previously never been ill, had never been in a hospital, and had never fractured a bone. He was a robust young man. On examination, the only findings apart from an initial fever, slightly accelerated pulse and breathing rate was that he was obviously sunburned on his face and neck. Curiously, there was also a pin point lesion on the palm of his left hand with no surrounding redness, warmth or swelling.

My fellow at the time, David Reagan, deserves full credit for asking for the details of his leisure time activities, how he acquired his sunburn, and exactly what he was doing in the week before his illness. A native of Tennessee, David is by nature very patient, engaging, soft-spoken and meticulous with details. People seem inclined to tell him everything, a gift for a clinician especially when dealing with patients who may not recognize the special importance of their routines or who may be reticent to review them initially.

The patient described a warm and sunny weekend four days prior to his hospital admission, when approximately 10 of his friends from rural Iowa gathered along the banks of a wide river with several cases of beer. They then organized a fire for cooking and began the major event, in reality a rite of passage in rural Iowa, hand fishing. Diving six to eight feet into murky brown water, they would blindly feel along the muddy river bed, and when they felt a bottom feeding fish, they would grab the gills with one hand and the tail with the other and rise to the top of the water with dinner.

When he was clasping his hand around one of the seven or eight large fish he successfully caught, he was spiked by the spine of a carp, but that injury was not recognized as a significant event. This may be because of the numbing effect of the alcohol imbibed or because hand fishing advocates ignore this commonly recognized hazard of the sport. When we later called experts with the State Health Department, we learned that Klebsiella was the most commonly found organism in the rivers of Iowa.

It is illegal to go hand fishing in Iowa, but it is a common practice in many communities. One authority told me that there is a low but occasional risk of dying from hand fishing. If a 40-50 pound fish bites a person who inadvertently placed his hand into the creature's mouth deep under water, it may be impossible to lift the fish up, and a hand fisher could drown. Our patient had heard of this but was undeterred with such tales.

I would learn that American Indians taught hand fishing to the French immigrants who had migrated from Nova Scotia to Southern Louisiana. The name of the French colony in Eastern Canada where they originally lived was called Acadia. Later called Cajuns, these people integrated hand fishing with festive social events that included music, story telling, and cooking by river banks. Presumably the tradition was commonly taught to settlers living nearby the sluggish stretches of big rivers on the Prairie.

One might ask, how do the hand fishers know what they are feeling? Apparently a catfish is smooth, has a flat head and can weigh 30-40 pounds commonly but can reach up to 100 pounds. In contrast, a carp is scaly. Neither the outline of a snapping turtle nor a snake poses any confusion even though such encounters in the deep offer new hazards. Of course I would never have had any insight into this American tradition if a few Gram negative rods from a river in Iowa had not penetrated the skin of a young man carrying on a tradition from Native Americans.

Our patient would receive intravenous antibiotics for three days, remain quite stable with resolution of his fever, and he would be discharged well on oral antibiotics for another week.

Still another opportunity came my way on a chilly fall night in Iowa. I had driven to the Cedar Rapids airport with my son Rich to meet my wife who was arriving on a plane from Chicago about 10:30 pm. When we met, JoGail told me that a middle-aged male passenger looked so ill that a woman next to him moved to my wife's row and sat down beside her, apparently fearing that the man's diseased condition might be contagious.

Looking back at him during the flight, my wife noticed that he was ghostly pale and sweating profusely. Although not thin he gave the appearance of being extremely weak and fragile.

I often tease JoGail when we travel because in contrast to my burying myself in work or reading like a recluse, she meets everyone within several aisles and takes a good social history. On the night of her arrival at Cedar Rapids she spoke to the ill appearing man after debarkation and offered my services if he would be interested.

He respectfully declined but asked if he could borrow some money from me for a call to Iowa City where his friend was awaiting his return to the country. When there was no answer, he returned and asked if I would give him a ride to Iowa City since that is where we too lived. I agreed to do this, and we placed all of our luggage in the car and drove off about 11:00

p.m. with the ill passenger in the front seat opposite me, and my wife and son in the back seat.

I asked him, "Where have you traveled to recently?"

"I have been living in Spain for a year's sabbatical" he replied.

"How long have you been ill?" I inquired.

He said that he had had "some fever for over a week, just a bad case of the flu." His muscles were achy, he was having some sweating periodically and also an impressive headache.

"May I ask what your occupation is, and by any chance did you travel to Africa at anytime while you were in Spain?" I asked.

"I am a Professor at the University, and I visited my daughter in Africa six weeks ago," he said.

"Did you notice any mosquito bites while you were there?" I asked. "Yes, on many occasions," he said. "Everyone gets bitten there."

"Did you take your malaria pills while you were there?" I followed up.

"No, I forgot to take them on several days," he admitted, "but this is just the flu."

I offered to take him to the hospital but he demurred, and after an hour looking for his friend, we unloaded the car of his luggage at his home and I gave him my card.

The next day he arrived at the hospital, showing the emergency room physician my card and reviewing our conversation the night before, The diagnosis of Vivax malaria - caused by *Plasmodium vivax* - was confirmed by a smear of his blood, which showed the characteristic ring-shaped parasites within the red blood cells. After appropriate therapy his health returned.

The association between the bad air of marshes – mal aria – and the illness called malaria has been recognized for thousands of years. We now know that four parasites of the species *Plasmodium* can enter the red blood cells of victims after the bite of the female anopheline mosquito (males cannot take blood meals) that breeds in marshes and other free-standing water. Although a rare infection in the U.S. now, there were a half a million cases each year here in the 17th and 18th centuries. Globally two million cases of malaria occur each year, mostly in Africa and Asia, and a few returning travelers can be harboring the parasite. Even those who take their pills religiously can acquire the disease because of increasing antibiotic resistance in the parasite.

The Romans recognized the cyclical fevers of malaria and differentiated tertian and quartian malaria with the fever accompanying the rupture of the red blood cells at every 48 and 72 hours, respectively. Although the designation may seem incorrect to us, recall that the Romans thought of the first day as day 1, thus 24 hours later would be day 2, and 48 hours later day 3 (tertian). *P. vivax* and *P. ovale* tend to be tertian, and *P. malariae* tends to be quartian in its cycle. *P. falciparium*, the most dangerous and life threatening form, can cause daily fever. My patient who arrived from Africa via Europe had returned home with additional baggage, one of the forms of tertian malaria.

He not only trivialized his trip to Africa to visit his daughter, his neglecting to take the antimalarial prophylaxis and his mosquito bites, but also his symptoms. Fortunately, he had the relatively benign form of malaria but one that can relapse. During his brief stay in the hospital he also received medications to preclude any relapses. He was recognized early because of my wife's caring personality at the airport and a brief inquiry into his social history during the drive from the airport to his home in Iowa City.

Fever is often the symptom uniting patient and physician. Earlier in my career at the University of Virginia I had the privilege of consulting on a young man who recently graduated from college and was referred to me because of fever. He was 21 years old, and on his way back from California he stopped in a city in Texas where he was given a tentative diagnosis of lymphoma because of large mediastinal lymph nodes noted on chest x-ray. These are the swollen glands that arise from the center of the chest near the heart and lungs.

Because he lived in Roanoke and had a relative on the faculty at the University of Virginia Medical School, he returned to Charlottesville both for a second opinion and a biopsy of one of the nodes within the chest. I was the attending physician on the infectious diseases service at that time.

We talked for awhile, and I learned that he had had a summer job in Bakersfield, California and surrounding cities. He was outside a great deal of the time and toward the end of the summer caught "the flu" just before travelling home. When he felt particularly ill in Texas, he worried about his health and sought consultation.

He could not recall any tick or mosquito bites; he had not traveled outside of California except to drive to and from there. He had no animal exposures, had not cared for any pets during his time there. In his free time, he was mostly outdoors with no unusual hobbies such as spelunking. He had no contacts with persons known to have tuberculosis.

Of course he could have had a lymphoma. The chest film was consistent with that diagnosis. But he had no symptoms of weight loss, fatigue or weakness in the weeks before his symptoms to suggest something more chronic. Nevertheless, the distinction between cancers, especially lymphomas, and infections, especially TB and fungal infections, is difficult.

Some geographic locations always trigger the possibility of an indigenous illness. One example are cities such as Bakersfield, California and others in arid climates at low altitudes and where the soil is alkaline. In the southwestern part of the United States these areas are along what is called the Lower Sonoran Life Zone. California, Arizona, New Mexico and parts of Texas would be included.

The soil in these areas is full of the spores of a fungus called *Coccidioides immitis*, the cause of a lung infection called coccidioidomycosis. When dust storms occur, epidemics of "Cocci" can occur afterwards. Newcomers to the Lower Sonoran Life Zone breathe in the spores and become infected with no recognized symptoms. Others become ill with a flu-like illness. A minority becomes more ill with progressive symptoms or signs of pneumonia. Because the white blood cells called lymphocytes that live within the mediastinal nodes – the glands that drain the infected portion of the lungs – attempt to manage the infection, the nodes swell from the inflammation in an analogous fashion to swollen glands in the neck after a Strep throat infection. On chest x-ray the swelling can be visualized in the center of the lung field.

The patient was scheduled to have a surgical procedure for a lymph node biopsy in 48 hours. In general, my examination was unhelpful in differentiating a curable fungus infection from a more formidable diagnosis of a malignant lymphoma, a cancer. However, he did have a small, five millimeter skin lesion that looked like a mosquito bite on the inside of his right arm. I asked him if it was recent and if he was sure that he had had no insect bite. It was not red, but flesh-colored and only slightly raised above the skin. He replied that he in fact had the small bump on his skin for a few weeks and could not recall an insect bite.

I immediately called my colleague in dermatology an asked him to biopsy the skin lesion. It was trivial in appearance, a "ditzel" in medical jargon, in other words appearing to be of little consequence. An excisional biopsy was done, and the results would be available early the next day.

I then went to the pathology text to see if the special stains for "Cocci" would be distinct on a biopsy specimen. Indeed they were. The somewhat spherical outlines of the fungus were quite characteristic and even to a non-pathologist seemed easy to spot. To me they had the shape of a bullet.

On the following morning I viewed the biopsy through the microscope with the dermatological pathologist, and we both saw the characteristic spherules: he had Cocci, not lymphoma! He had a fungal infection. And because we had the diagnosis from a simple biopsy of the skin, he would not need the more invasive lymph node biopsy of his mediastinum. A good social history, a careful physical examination, and an aggressive approach for a skin biopsy combined to give the good news.

After weeks on an intravenous antibiotic called Amphotericin B he was completely cured.

Coccidioidomycosis, first described in a patient in 1892, was erroneously thought to be a parasite initially, a member of the genus *Coccidia*, and because it was serious - not mild – the species was called "immitis." The spores can become airborne and ride the waves of the winds, as on magic carpets, eventually dispersing the organism for tens of miles before settling on communities of people below. Thus, the initial infection involves the lungs when people living in endemic areas breathe in the fungus.

For years this disease was called "desert fever" because of its association with the areas in the U.S. Southwest. Current estimates are that 100,000 people in the U.S. are infected each year, sometimes in travelers who have spent only hours or days in the endemic area. Only a small proportion of all those infected go on to some degree of illness, the result of the body's failure to contain the fungus. My own patient represented one of the exceptions.

Sometimes the new students on the wards and even some of the residents will say, "the only part of the social history relates to how many drinks a patient consumes each day, how many cigarettes smoked, and the use or non-use of illicit drugs." This is short sighted and will lead to a series of delayed diagnoses or misdiagnoses and possibly to unnecessary diagnostic procedures if not corrected. Of course our job as mentors is to show the way and illuminate the importance of a careful social history. Some narrative stories help.

EPILOGUE

Medicine is an art, but it has a strong discipline involving the elements of the history, the physical examination, laboratory testing and follow up evaluations. All parts of the history including the social history are essential, and in infectious diseases the social history cannot be minimized. When a key disease is identified that pinpoints a likely diagnosis, there is enormous excitement for the clinician – another mystery solved. This has always been an attraction for me and many in the field of medicine. I think for young people interested in a future in medicine, it is a strong motivator. Furthermore, within the youngest science – medicine – lies the specialty of infectious diseases with its rich history, acute challenges, its mystery and many rewards.

The insightful microbiologist and author well known in the early 20th century, Hans Zinsser, described much of my own sense of romance in my specialty: "Infectious disease is one of the few genuine adventures left in the world. The dragons are all dead, and the lance grows rusty in the chimney corner".

My experiences in medicine and infectious diseases furthermore have helped me reflect on several questions: What is a physician and how should a doctor practice his or her science and art? In a 1927 article in *JAMA* the thoughtful clinician, Francis Weld Peabody, crystallized what my career has led me to consider. He said, "The secret of the care of the patient is in caring for the patient". Throughout my career I have periodically reflected on the topography of our field, the face of medicine. Up to date knowledge is essential, but where are creativity and rational decision making? Where are ethics, high standards, responsiveness, and tenacity to be found? Where is wisdom? I have suggested that an innate curiosity and independent thinking are especially important. The interactions of man and microbe, distant cousins on the evolutionary tree, are perennial features of life on earth. Although it is in the best of interests of the organisms not to overwhelm the host, loss of health and sometimes death can result from what *Homo sapiens* call infection. A physician's role is to intercede on behalf of individual people and sometimes populations to restore comfort and if possible to preserve life itself. To do this well and efficiently, the doctor has to be a caring individual, curious and competitive with microbes in seeking exact diagnoses and treatments, and focused on the interests of those putting their deepest trust in us.

Can one imagine what tomorrow will bring? Very likely more mysteries, puzzles and surprises – dispensed with large doses of novelty and uncertainty that characterize the practice of Medicine. However, some obvious clues to the new challenges are available. I would address three for consideration:

1) New Infections
2) Increasing Understanding of Science
3) Increasing Technology in the Care of Patients

In the Spring of 2003, we faced the scary news in daily headlines of an emerging epidemic called SARS, Severe Acute Respiratory Syndrome. The causative virus is a novel coronavirus, possibly originating from a civet cat – considered a culinary delicacy in the southern part of China in the fall of 2002. Perhaps the proximity of man and animals – especially among chefs and animal handlers – led to the microbial accidental tourist crossing species lines. Thereafter, transmission to a susceptible global population was enhanced by available international travel. Within hospitals, health care workers in close proximity to SARS patients became ill, some succumbing to the novel pathogen.

The resulting decline in national and international travel had enormous economic consequence for countries in Asia and later for Canada, and the increasing fear of infection and death created daily anxieties. For the first time in decades, physicians and nurses worried about their own vulnerabilities, even their health. Some worried about dying if infected because the case fatality was estimated to be 9%. However, for those older than age 60, half of the victims died.

This is only the latest of easily 30 new infections identified in the past 25 years, joining earlier ones such as the causes of toxic shock syndrome (staph and strept), Legionnaire's disease, hantavirus pulmonary infection in the four corners region of the Southwest U.S., Ebola, HIV infection, West Nile virus and monkeypox in the U.S. and others. This is the history of man and microbe, yet we always seem to be astonished with each new pathogen greeting us. Like wars and natural disasters, new epidemics of infection occur regularly. Sometimes they skip generations and are perceived as novel and improbable, when in fact they are ageless and routine.

There are several lessons from such experiences. We must maintain an interest in whatever occurs in remote areas of the world. An initial conviction that disease we read about in other people oceans away from us has no relevance for us usually falters with time. Second, many changes in the ecosystem or major social and lifestyle changes can have important and unexpected consequences. The rise in the deer mouse populations the year of the outbreak of hantavirus among native Americans living in Arizona, Colorado, New Mexico and Utah in 1993 led to the epidemic in people. Gay lifestyles initially amplified the infection due to HIV virus, and unprotected sex among heterosexual couples has maintained the epidemic in many parts of the world.

Globalization allows the sale of food from one country to another with occasional outbreaks related to contaminated meat products (Mad Cow Disease), cheese (Listeria bacterial infections especially of pregnant women and newborns), and fruit and vegetables (various parasites of the gastrointestinal tract). And SARS is just the most recent epidemic amplified by airline travel in brief period of time.

A second consideration in thinking about the future of medicine relates to the explosion of science with translation of scientific discoveries from the bench in the laboratory to the bedside. With the information derived from the human genome project, in which every gene will be identified, there will be a time when we can draw blood and give patients probabilities of acquiring various diseases including specific infections. In addition, our genes make us not only at higher or lower risk of infection but also code for our response to infection, whether or not we develop shock, or even die.

Of course new issues arise with such abilities to "type" people: will everyone want to know their risks? Most will if there is a safe prevention, but what if no cure or prevention is available? Will people use the information in the selection of mates? Will insurance companies or employers try to acquire the information and use it to screen applicants? Would medical school, the military or other professional guilds try to use the information in their application process? Much of this is to say that science often outpaces the ethical considerations of such advances.

A third consideration relates to the use of technology in the care of patients. Here I am particularly focusing on devices and procedures rather than technological advances in medications. New, less invasive surgery should continue to be developed with results to decrease currently expected morbidity. Abscesses can increasingly be drained by radiological guidance techniques rather than by a major surgical procedure. Electrical devices can be placed to stimulate the heart to beat at a set rate, to shock the heart if it goes into an aberrant and dangerous rhythm, or to help it pump blood more effectively. We might expect newer devices to help breathing efficiencies, ambulation, voice control and others. Our ICUs will increasing pursue the use of life sustaining machinery.

Each advance in technology, however, has a new and often unexpected risk of new infections. Initially these risks might be very high as was experienced with external heart pumps assisting the patients' own heart muscle. With time, improved protocols for insertion and management reduced the risks.

An even more challenging problem with devices is this: technology forms barriers between patients and their physicians and nurses. Some students or physicians may be heard to refer to "the man in room 34 with the pacemaker" or "the woman in the ICU on a respirator." Or if the external heart assist device (electrical pump) develops any malfunction, *it* may become the focus of attention, not the patient, who is feeling his or her life at risk with such malfunctioning.

What I am worried about is that technology will widen the expanding gulf between patients and their physicians, will perturb the doctor-patient relationship in a negative fashion. This time-honored, hallowed and sacred bond between two people is essential for care and for cure when possible. Patients have to feel free to ask any question, to report any sign or symptom, to express their deepest fears and long held secrets. None of this will occur if we lose the doctor-patient relationship as a result of expanding modern technology.

At the time of this writing I have no simple solution. But I will continue to teach the reverence we must have for our patients. I will continue to stress the ideals of our profession and the enormous rewards if we gain the trust of those who initially greet us in our offices and hospitals as strangers with symptoms of illness.

READINGS

The Prince of Pathogens
(Staphylococcus aureus)

Adam A. Alexander Ogston and the Army Medical Services Formation of the Royal Army Medical Corp 1 July 1898. *Scot Med J.* 1998; 43: 156-7

Alexander Fleming. The Man and The Myth. Gwyn MacFarlane. Howard University Press, Cambridge, MA, 1984

Edmond MB, Wenzel RP, Pasculle AW. Vancomycin-Resistant *Staphylococcus aureus*: Perspectives on Measures Needed for Control. *Ann Intern Med.* 1996; 124: 329-334

Fleming A. [Classics in Infectious Diseases] On the Antibacterial Action of Cultures of a Penicillium, with Special reference to their use in the isolation of B. Influenzae. Rev Inf Dis 1980; 2: 129-139

Hollis RJ, Barr J, Doebbeling BN, Pfaller MA, Wenzel RP. Familial Carriage of Methicillin-Resistant *Staphylococcus Aureus* and Subsequent Infection in a Newborn Sibling. *Clin Infect Dis* 1995; 21: 328-332

Lyell A. Alexander Ogston, Micrococci, and Joseph Lister. Epitaph. *J. Amer. Acad Dermatol.* 1989; 20: 302-310

M.D. One Doctor's Adventures Among The Famous and Infamous from The Jungles of Panama To a Park Avenue Practice. B.H. Kean, MD with Tracy Dahlby. Ballantine Books. New York., 1990

Ogston A. "On Abscesses". Classics in Infectious Diseases. *Rev Infect Dis.* 1984; 6: 122-128

Peacock JE, Marsik FJ, Wenzel RP. Methicillin-Resistant *Staphylococcus aureus*: Introduction and spread within a hospital. *Ann Intern Med.* 1980; 93:526-532

Perl TM, Cullen J, Wenzel RP, et al. Intranasal Mupirocin to Prevent Postoperative *S. aureus infections. N Engl J Med* 2002; 346: 1871-7.

Reagan DR, Doebbeling BN, Pfaller MA, Sheetz CT, Houston AK, Hollis RJ, Wenzel RP. Elimination of Coincident *Staphylococcus aureus* nasal and hand carriage with intranasal application of mupirocin calcium ointment. *Ann Intern Med.* 1991; 114: 101-106.

Sanford MD, Wider AF, Bale MJ, Jones RN, Wenzel RP. Efficient Detection and Long Term Persistence of Methicillin-Resistant *Staphylococcus Aureus. Clin Infect Dis* 1994; 19: 1123-1128

Schneeberg NG. Death of a President. *N. Engl J. Med.* 2000; 342: 1222

Skinner D, Keefer CS. Significance of Bacteremia Caused by Staphylococcus Aureus. *Arch Intern Med* 1941; 68:851-75

Wenzel RP. The Antibiotic Pipeline – Challenges, Costs and Values. *N Engl J Med* 2004; 351:523-6

Woods S, Edwards S. Philosophy and Health. *J. Adv. Nurs* 1989; 14:661-4

.

K.G.
(Vibrio cholerae)

Love AH, Phillips RA Measurement of Dehydration in Cholera. *J. Infect Dis*. 1969; 119: 39-42

Phillips RA. Twenty Years of Cholera Research. *JAMA* 1967; 202: 610-614

Phillips RA, Van Slyke DD, Hamilton PB et. al. Measurement of Specific Gravities of Whole Blood and Plasma by Standard Copper Sulfate Solutions. *J. Biol Chem* 1950; 183: 305-330.

Savarino SJ. A Legacy in 20th Century Medicine: Robert Allen Phillips and the Taming of Cholera. *Clin Infect Dis* 2002; 35:713-20

Wenzel RP, Phillips RA. Intraperitioneal Infusions for Initial Therapy of Cholera. *Lancet* 1971; 2: 494-5

The "Corps"
(Mycoplasma pneumoniae)

Chanock RM. Mycoplasma Infections of Man. *N. Engl. J. Med.* 1965 23:1257-1264/ 22: 1199-1206

Chanock RM, Hayflick L, Barile MF. Growth on Artificial Medium of Agent Associated with Atypical Pneumonia and its Identification as PPLO. *Proc Nat Acad Sci.* 1962; 48: 41-49.

Chanock RM, Dienes L, Eaton MD, et al Mycoplasma pneumonia: Proposed Nomenclature for Atypical Pneumonia Organism (Eaton Agent), Science 1963; 140: 662

Perkins JC, Tucker DN, Knopf HLS, Wenzel RP, Hornick RB, Kapikian AZ, Chanock RM. Evidence for Protective Effect of an inactivated rhinovirus vaccine administered by the nasal route. *Am J Epidemiol* 1969; 90:319-326.

Perkins JC, Tucker, DN, Knopf HLS, Wenzel RP, Kapikian AZ, Chanock RM. Comparison of Protective Effect of neutralization antibody in serum and nasal secretions in experimental rhinovirus type 13 illness. *Am J Epidemiol* 1969; 90:519-526.

Wenzel RP, Craven RB, Davies JA, Hendley JO, Hamory BH, Gwaltney JM JR, Field Trial of an Inactivated *Mycoplasma Pneumoniae* Vaccine. I. Vaccine Efficacy. *J. Infect Dis* 1976; 134: 571-576.

Wenzel RP, Hendley JO, Dodd WK, Gwaltney JM Jr. Comparison of josamycin and erthromycin in the therapy of *Mycoplasma pneumoniae* pneumonia. *Antimicrob Agents Chemother* 1976; 10:899-901

| Discovery |
| (Rickettsia) |

Beatty WK, Beatty VL. Howard Taylor Ricketts: Imaginative Investigator. *Proc Inst. Med. Chgo* 1981; 34: 46-48.

Hechemy KE, Fox JA, Groschel DAM, Hayden FG, Wenzel RP. Immunoblot studies to analyze antibody to *Rickettsia typhi* group antigen in sera from patients with acute febrile cerebrovasculitis. *J. Clin Microbiol* 1991; 29: 2559-2565.

Linnemann CC Jr, Peetzman CI, Peterson ED. Acute Febrile Cerebrovasculitis. A non-spotted fever group Rickettsia disease. *Arch Intern Med.* 1989; 149: 1682-1684

McNee JW, Renshaw A, Brunt EH. "Trench Fever". A Relapsing Fever Occurring with the British Forces in France. *Brit Med J.* 1916 (Feb 12): 225-274.

Weiss E, Strauss BS. The Life and Career of Howard Taylor Ricketts. *Rev Infect Dis.* 1991; 13: 1241-1242.

Wenzel RP, Hayden FG, Groschel DHM, Salata RA, Young S, Greenlee JE, Newman S, Miller PJ, Hechemy KE, Burgdorfer W, Peacock JG. Acute Febrile Cerebrovasculitis: A Syndrome of Unknown, Perhaps Rickettsial, Cause *Ann Intern Med.* 1986; 104: 606-615.

Deaths in the ICU
(Serratia marcescens)

Donowitz LG, Marsik FJ, Hoyt JW, Wenzel RP. *Serratia Marcescens* Bacteremia from Contaminated Pressure Transducers. *JAMA* 1979; 242: 1749-51.

Firkin B.G, Whitworth J.A. Dictionary of Medical Eponyms. The Parthenon Publishing Group. *Park Ridge N.J.* 1987.

Greenberg L. Serratia Marcescens in Human Affairs *Drug Intelligence Clin Pharm* 1978; 12: 674-679

Sehdev PS, Donnenberg MS. Bartolomeo Bizio. *Clin Infect Dis.* 1999; 29:925

Wenzel RP, Osterman CA, Hunting KT, Gwaltney JM Jr. Hospital-Acquired Infections I. Surveillance in a University Hospital *Am J. Epidemiol* 1976; 103: 251-260

Yu VL, Serratia Marcescens. Historical Perspective and Clinical Review. *N. Engl J. Med.* 1979; 300: 887-893

In the Shadow of Semmelweis
(Enterococcus, Proteus mirabilis, Streptococcus pyogenes)

Bischoff WE, Reynolds TM, Sessler CN, Edmond MB, Wenzel RP. Handwashing Compliance by Health Care Workers. The Impact of Introducing an Accessible, Alcohol-Based Hand Antiseptic. *Arch Intern Med* 2000; 160: 1017-21.

Doebbeling BN, Stanley GL, Sheetz CT, Pfaller MA, Houston AK, Annis L, Wenzel RP. Comparative Efficacy of Alternative Handwashing Agents in Reducing Nosocomial Infections in Intensive Care Units. *N. Engl J. Med.* 1992; 327: 88-93

Edmond MB, Wallace SE, McClish DK, Pfaller MA, Jones RN, Wenzel RP. Nosocomial Bloodstream Infections in United States Hospitals: A three-year analysis. *Clin Infect Dis.* 1999; 29: 239-234.

Wenzel RP, Pfaller MA. Handwashing: Efficacy vs Acceptance. A Brief Essay. *J. Hosp. Infect* 1991; 18 (Suppl B) : 65-68

Green JW, Wenzel RP. Postoperative wound infections; A controlled study of the increased duration of hospital stay and direct cost of hospitalization. *Ann Surg* 1977;p 185:264-268.

Morrison AJ, Freer CV, Searcy MA, Landry SM, Wenzel RP. Nosocomial Bloodstream Infections: Secular Trends in a Statewide Surveillance Program in Virginia. *Infect. Control* 1986; 7: 550-553.

Wenzel RP, Edmond MB. Managing Antibiotic Resistance. *N Engl J Med* 2000; 343: 1961-7

Wenzel RP, Thompson RL, Landry SM, Russel BS, Miller PJ, Ponce de Leon S, Miller GB. Hospital-Acquired Infections in Intensive Care Units: An Overview with Emphasis on Epidemics. *Infect Control* 1983; 4: 371-375

Wenzel RP, Osterman CA, Townsend TR, Veazey JM Jr, Servis KH, Miller LS, Craven RB, Miller GB, Jackson PS. A Statewide Program for Surveillance and Reporting of Hospital-Acquired Infections. *J Infect Dis* 1979; 140:741-746

Wong MT, Kaufmann CA, Sandiford HC, Fuchs HJ, Wenzel RP and the Ramoplanin-VRE 2 Clinical Study Group. Effective Suppression of Vancomycin-Resistant *Enterococcus* in Asymptomatic Gastrointestinal Carriers by a Novel Glycolipodepsipeptide, Ramoplanin. *Clin Inf Dis* 2001; 33: 1476-82.

> **The Origin of Fear**
> (Neisseria meningitidis)

Edmond MB, Hollis RJ, Houston AK, and Wenzel RP. Molecular Epidemiology of an Outbreak of Meningococcal Disease in a University Community *J. Clin Micro* 1995; 33: 2209-11

Scholz A, Wasik F, Albert Neisser, 1855-1916. *Internat J Dermatol* 1985; 24 (July-Aug): 373-7

Wenzel RP, Davies JA, Mitzel JR, Beam WE Jr. Non-usefulness of Meningococcal carriage-rates. *Lancet* 1973;2:205

A Bump in the Night
(Bacteroides, Klebsiella pneumoniae, Plasmodium vivax, Coccidioides immitis)\

Reagan DR, Nafziger DA, Wenzel RP. Handfishing – Associated Klebsiella Bloodstream Infection. *J. Infect Dis.* 1990; 161: 155-156.

Woodward TE. The Golden Era of Microbiology: People and Events of the 1880s *Maryland Med J* 1989; 38:323-328.

INDEX

A

Acadia 115
Athens 26

B

Bacteroides 86, 111, 112, 113
Baltimore 25, 28, 30, 33, 37, 100
Bangladesh 28, 30
Bay of Bengal 25
Beaufort, South Carolina 40, 41
Bloodstream Infection 131, 133
Brucella 57
Bubo virginianus 113
Budapest 89, 90

C

Cajuns 115
Campylobacter 17, 75
Camp LeJeune 34, 37, 38, 40, 42,
    44
Cam Ranh Bay 33
Centers for Disease Control 67,
    71, 88
Charlottesville 40, 51, 55, 60, 65,
    100, 117
Chestnut Hill Academy 8
Chlamydia 111
Copenhagen 56
Copper Sulfate Solutions 127
Critical Care 65, 68
CT 52, 54, 55, 125, 131
Cytomegalovirus 53

D

Dacca 25, 26, 27, 28
Da Nang 33
Dehydration 127
DNA 16, 84, 85, 105, 106

Duke University 35

E

E. coli 15
East Pakistan 25
Ebola 122
EKG 28
Elek and Cohen 4

F

FDA 12, 74
Field Research Laboratory 37
Flat Head Indians 59
Food and Drug Administration 74
Fripp Island 41, 43

G

Ganges River 27
GlaxoSmithKlein 12
Gram Negative 86

H

Haverford 8, 16, 36
Herpes simplex 53
Histoplasmosis 111
HIV 58, 60, 122, 123
Homo sapiens 12, 121
Hospital Epidemiologist 66
Human Immunodeficiency Virus
    58

I

ICU viii, 51, 54, 65, 66, 68, 69,
    70, 71, 72, 73, 74, 76, 77,
    94, 95, 96, 100, 104, 124
ID Rounds 104
Immunoblot 129
Inaba 19

Internal Medicine vii, 18, 25, 31, 33, 40, 66, 96
IV 20, 21, 23, 27, 29, 65, 70, 71, 76, 87, 91

J

Jefferson Medical College 8, 15
Johns Hopkins 62, 100

K

Karachi 26
Kirschner wire 3, 4, 8
Klebsiella pneumoniae 56, 111, 113
Kolechka 90

L

Lin Yutang 23
Luzon 23

M

M. tuberculosis 113
Madagascar 17
Malaysia 17
Manila 20, 21, 22, 23, 24, 25, 26
Marine Corps 15, 34, 37, 38, 39, 40, 41, 44, 45
Mekong Delta 33
Merck 12
Methicillin 125
Mexico 59, 60, 118, 122
Morocco 17
MRI 89, 90
Mycoplasma pneumoniae 33, 34, 40, 44, 53, 128

N

National Institute of Allergy and Infectious Diseases
   s 33

NIH 33, 35, 36, 40, 43
Nosocomial 67, 77, 131

O

Ogawa 19

P

P. malariae 117
P. vivax 117
Parris Island 34, 39, 40, 41, 43, 44
Penicillin 9, 104
Pfizer 12
Philadelphia 3, 8, 15, 20, 22, 24, 33, 69
Philadelphia General Hospital 24
Philippines 20, 22, 26
Plasmodium vivax 111, 116
Proteus 81, 82, 92
Pseudomonas aeruginosa 91, 92

R

Rhinovirus 6
Richmond viii, 84, 88
Rickettsia 51, 57, 60, 129
Rickettsia prowazekii 57
Rickettsia typhi 57, 60, 129
Rocky Mountain Spotted Fever 53, 57, 59, 60

S

S. epidermidis 112
Salmonella 15, 17, 84, 111
San Lazaro Hospital 20, 22
SARS 122, 123
Serratia marcescens 65, 68, 70, 72, 74, 75, 76
Severe Acute Respiratory Syndrome 122
Shigella 15, 17, 84
Sick Bay 40

Six Day War 26
Skinner and Keefer 9
Southeast Asia Treaty Organiza-
    tion 25
Staphylococcus aureus 3, 4, 6, 7,
    9, 11, 125
Streptococci 4
Streptococcus pyogenes 81
Striges 112, 113
Surveillance 83, 87, 88, 130, 131

T

Taipei 18, 19, 23, 24, 25
Taiwan 19, 23
The New England Journal of
    Medicine 6, 94
Tularemia 45, 57
Typhoid 45
Typhus 58, 59

U

University of Iowa 6, 93, 99, 100,
    102, 104, 105, 111, 113
University of Pennsylvania 60
University of Virginia 40, 51, 65,
    117

V

Vaccine 128
Vancomycin 84, 125, 131
Vibrio cholerae 15, 17, 19, 127
Vienna 89, 95
Viet Nam 33, 37, 46, 99
Virulence 83
Vivax malaria 116

W

Woods and Edwards 11

ABOUT THE AUTHOR

Dr. Richard P. Wenzel is Professor and Chairman of the Department of Internal Medicine at Virginia Commonwealth University. He is recognized by many as one of the leading infectious diseases epidemiologists in the United States.

Dr. Wenzel was named one of ten contemporary "Great Teachers" by the National Institutes of Health in their 2001-2 series. He is a prolific author of over 450 publications and editor of six textbooks on Infection Control and Quality Health Care. He has won numerous awards for research and in 2001 was named Editor-at-Large for *The New England Journal of Medicine*. He is President-Elect of the International Society for Infectious Diseases.

He and his wife JoGail live in Richmond.

Printed in the United States
31068LVS00002B/100-126

9 781420 820065